Speech in the User Interface:
Lessons from Experience

William Meisel
(Editor)

Published by TMA Associates (www.tmaa.com, info@tmaa.com)

Order this book online at www.trafford.com
or email orders@trafford.com

Most Trafford titles are also available at major online book retailers.

Printed in Victoria, BC, Canada.

ISBN: 978-1-4269-2622-8 (sc)

Our mission is to efficiently provide the world's finest, most comprehensive book publishing service, enabling every author to experience success. To find out how to publish your book, your way, and have it available worldwide, visit us online at www.trafford.com

Trafford rev. 1/18/2010

 www.trafford.com

North America & international
toll-free: 1 888 232 4444 (USA & Canada)
phone: 250 383 6864 ♦ fax: 812 355 4082

Contents

Speech in the User Interface: Lessons from Experience

Preface

This collection of articles is intended to aid the design of user interfaces to applications using speech recognition, text-to-speech, speaker verification, audio search, and other speech technologies, often integrating speech with other modes such as text display or touch screens. The articles in this collection were written by experts experienced in delivering commercial applications and gathering insights from user reactions. The articles, other than the editor's, were originally published as guest columns in the "VUI Visions" (VUI = Voice User Interface) section of the newsletter *Speech Strategy News*. The chapters written by the editor, Bill Meisel, were written for this collection to bring together and expand editorials from the newsletter.

About the editor

William "Bill" Meisel is president of TMA Associates, a speech-industry consulting and publishing firm (www.tmaa.com). Meisel writes the newsletter *Speech Strategy News* (since 1993), is co-organizer of the Mobile Voice Conference (www.mobilevoiceconference.com), Executive Director of the non-profit Applied Voice Input Output Society, and edited an earlier volume in this series in 2006, *VUI Visions: Expert Views on Effective Voice User Interface Design*. In the 1980s, he founded and ran a speech recognition technology company that did early work in automating customer service and continuous-speech dictation of medical reports, after managing the Computer Science Division of an engineering firm. Meisel began his career as a university professor, wrote the first textbook on Computer Pattern Recognition, and has published numerous technical papers. He holds a B.S. in Engineering from Caltech (California Institute of Technology) and a Ph.D. in Electrical Engineering from USC (University of Southern California).

Introduction

William Meisel

Innovations in user interfaces have been instrumental in technology advances and commercial successes. The most obvious example is the Graphical User Interface, today's face of personal computers, Web browsers, and, increasingly, mobile phones.

Interacting with machines by speech is a common science fiction theme, so it is not surprising that it has been seen as a possibility for commercial applications, perhaps the next major innovation in user interfaces. Unfortunately, the science fiction image of machines with human intelligence (or perhaps alien malevolence) has led to high expectations for systems using speech recognition and speech synthesis, expectations of human-like capabilities. Disappointment with the limitations of early speech technology led to skepticism that still exists today in many quarters, despite steady advances in the technology.

Interactions that reflect the full range of human dialog are still well out of sight, but speech recognition today can match or even exceed human capabilities in specific contexts. Consider, for example, automated business directory assistance delivered over the telephone. Today's systems may ask for city and state before a business name, but one can say any city-state combination in the country and expect relatively good accuracy. Similarly, stock quote systems typically do quite well as long as the company name is a listed stock. These types of applications can out-do a human in speed and accuracy, in part because they have a computer memory that remembers all the city-state combinations or stocks—unlike a human agent that can't know all the street or stock names and yet must understand them and type them into database software, then read the result to the caller. Classically, this is the type of job where automation is most effective—repetitive, tiring jobs better suited for a machine.

Large-vocabulary speech recognition, sometimes called "speech-to-text," is designed to create a transcript of spoken material. It is used on PCs by many who don't want to type or can't. A major application is medical reporting, where the limited context can increase recognition accuracy and contribute to complete and timely medical records. One vendor recently cited word accuracy as high as 99% for its radiology dictation solution. Speech-to-text software uses context through statistical models to increase accuracy, but doesn't attempt to "understand" what it is transcribing.

Another large-vocabulary application growing in popularity is delivering voicemail as text. This application has an advantage in that the recipient can often get the gist of the message even with errors, so a less accurate transcription than that required for dictation applications may suffice.

Similarly, text-to-speech (TTS) synthesis today is clearly intelligible and can read any text (with occasional mispronunciations and misplaced stress). In a typical application, information accessed from a database in text form can be spoken over the telephone. The most recent TTS technology connects slices of speech created from recorded speech ("concatenative" TTS), and sounds like a human that recorded the speech rather than robotic voice.

Speaker verification (speaker authentication) matches a person's voice to features of their voice (gathered during a verified enrollment) that are characteristic of the vocal tract. These "biometric" physical characteristics, sometimes called a "voiceprint," are unlikely to match an imposter. A fingerprint might be more secure, but voice can be delivered over the telephone and make fraud essentially impossible, certainly more so than typical information used such as one's mother's maiden name. Because speaker authentication can speed transactions that require verification and doesn't require special equipment, its use will continue to grow.

Speaker authentication can be integrated naturally and seamlessly with a more general speech interaction in a call center. Even if an attempt at fraud uses speech recorded from a tapped line, that will not work with a system that varies the required prompt, e.g., asking for one of various passwords or number combinations. In practice, a fraud attempt on an account protected by a speaker authentication is unlikely—there are much easier ways to commit fraud. With the increasing use of mobile smartphones that can include personal information, we can also expect the technology to eventually be ubiquitous on the phone itself to control data access.

Audio search—using speech recognition technology to find specific phrases in audio files, e.g., webcasts, forensic data, or recorded call center calls, can make recorded audio accessible by content. This is obviously much more efficient than a human listening to many hours of audio or video, and is another one of those cases where the task should be automated rather than tie up a person in an onerous job.

One aspect of audio search is "speech analytics," where the audio search is used to find audio files with specific characteristics and the point in those files where specific words signaling the category are found. Speech analytics has become particularly important in call centers to find out what is driving call volume or customer service problems.

"Voice search" can be considered one form of speech recognition. It is sometimes distinguished because it has a special use. The term is most commonly used when someone speaks rather than types a search request, often talking to enter text in a text search box on a mobile phone. The same "say what you want" paradigm can often be extended by using control words to distinguish types of requests, e.g., "search pizza" versus "call Dave." Applications that work this way have been deployed by a number of companies.

But suppose we grant that the technology works well enough today to support very useful applications. That doesn't mean it can't be done poorly. The classical example is some of the terrible web sites when web browsers were in their infancy (and perhaps a few today). Speech interfaces can be done poorly, and many certainly are. Some of the problems are caused by business decisions; in contact centers, for example, managers often decide to use speech technology on one function at a time, inadvertently preserving the clumsiness of an introductory and often confusing menu of choices.

The intent of this book is to provide experienced insights into what works and what doesn't. Hopefully, the hard-earned knowledge represented here will inform a new generation of experts, allowing them to avoid learning these lessons the hard way. At the same time, reference to documented good practices may help both experts and novices point out to business managers when they are making decisions at the expense of usability.

Organization of this book

The following three chapters, by the editor, are intended to set the context for more detailed discussions of best practices and issues in user interface design incorporating speech technology. The first puts speech interfaces into the context of the evolution of user interfaces and addresses what a universal voice user interface model might be. The second discusses trends and motivations for using speech interfaces in telephone and mobile applications. The third talks about the diverse applications in markets other than telephony.

Other chapters focus on drilling down into issues and solutions in more specific cases. An attempt has been made to cluster chapters with related content, but the principles discussed in most chapters often have relevance beyond the specific examples discussed. Don't expect a textbook—we've chosen authors with varying opinions, and you may rightfully sense some controversy on the right way to handle some issues. The only constant is that these authors are speaking from deep experience. Another constant I have observed is the strong commitment of the authors to discover what works.

Chapters are intended to be independent. A reader can peruse the Table of Contents and start at any chapter of interest.

The evolution of user interfaces in computing and communications systems

William Meisel

The Introduction alluded to the commercial success of innovative user interfaces. The Graphical User Interface (GUI) has certainly been a successful innovation, with the WIMP paradigm [Windows, Icons, Menus, Pointing device (e.g., a mouse, finger, or an arrow key)] driving PC operating systems and some mobile devices such as the iPhone. A brief review of the history of successful user interfaces may provide some insight into their future, yielding a perspective on the role of speech technology.

To a large extent, GUIs on PCs replaced 'command lines' as a user interface. Before PCs, mainframes and minicomputers were accessed through "dumb" terminals. One interacted with those operating systems by giving them typed commands that had to have a very specific structure and typically used non-English commands that one learned from a manual. Each command had a rigid set of parameters that had to follow the command. The approach guaranteed job security for the experts that managed to learn the command language, usually programmers. The first personal computers used a similar user interface.

There was some interactivity in command-line interfaces—the computer at least told you when the command didn't parse. This wasn't a particularly bad interface for those who were willing to use reference tomes and put in the many hours of study to master it. What was good about this interface—and we will return to this point later—was that there was no "navigation" among folders or scrolling through menus. You just told the computer what to do, and—if you spoke its language—it did it. The result was often a list of information that led to further command-line inquiries. But the command-line model required more and more study as applications multiplied, and its rigid syntax could not support mass adoption of computers.

The GUI WIMP interface drove mass adoption of personal computers and the growth of the Internet through graphical Web browsers. The "desktop" metaphor, a specific aspect of most WIMP interfaces, also works. We can spread out folders on our desktop, "open" them, and search through them. Making the options more visually obvious through an intuitive GUI reduced the need for a manual, and, equally importantly, led to some informal standards that made it easier to adapt to a new application. Every application seems to have a "file" menu with "open," "save," and "close" commands, for example. The desktop metaphor extends this model by adding Folders as a specific icon and the concept of a window that is the default starting window; our desktop is always there, albeit cluttered.

There is a lesson here. The GUI was successful because aspects of successful designs were copied, and the consistency was important in avoiding a steep learning curve for each application. There has not been much consistency in speech applications so far, perhaps accounting in part for the slow adoption of speech in user interfaces.

Exponential growth of the Web may have led to one of the most successful user interface innovations beyond the WIMP model—basic Web search by Google, Microsoft, Yahoo and others. Basic Web search is text-based: type in a search term and then see a list. A bit like a command line, isn't it? A key difference between classical command lines and search is that the format isn't constrained, and there is a lot of computation and technology behind understanding the user's intent as well as their words. The search text box may not seem like a "user interface" when compared to the complexity of today's GUI interfaces, but simplicity is a virtue in a user interface, not a fault.

We also use the "search" user interface on our PC to find a file we need. In both the web and the PC cases, the result of a search is a list. The WIMP interface aids the search paradigm by providing a means to select an item on the list, usually with a pointing device. In this sense, the simplicity of the search text box is possible because of its leveraging the existing GUI interface to deliver its results.

Again, there is a possible lesson here for speech interfaces. When another interface mode is available and can make a speech interface more effective, use it.

On the telecommunications front, the "user interface" to computing systems began as a human agent. The agent would talk to the customer, access information on their computer terminal, and provide information or make changes through oral discussion with the customer. This expensive model is still considered by many the preferred model, despite the fact that, today, most of the time we are using the extremely inefficient process of sticking an agent between a user and a computer to access a computer database, requiring the agent to retype what the caller says, often asking for the information to be repeated or spelled. The advantage of this approach to the caller is that it is a form of command line that accepts English statements, to perhaps stretch an analogy.

Since automation provides significant savings and speech recognition didn't work very well until fairly recently, touch-tone automation evolved as a limited option. Widely hated by callers, touch-tone has very little flexibility; it forces speaking a list of options that are either very long and annoying or short and confusing. Today, speech recognition provides a practical alternative in many cases, allowing agents to handle the truly complex cases that require human intelligence, not simply inefficient human simulation of a speech recognition system accessing a database.

Re-visiting the command line

And perhaps the command line hasn't quite gone away. Today's "power users" often bypass the complex navigation of WIMP interfaces by writing scripts (short programs) that can be launched, in effect, with a single typed command, yet cause a number of actions to be sequentially taken. One doesn't have to be a computer science major to use the scripting/programming capability in programs like Microsoft Excel.

And some shortcuts are getting built into programs, such as Microsoft Word's "Work" menu, which allows jumping quickly to documents that one uses frequently. As the WIMP interface gets increasingly overwhelmed with overlong menus and too many icons, folders, documents, and options, one is tempted to seek these shortcuts.

And one could ask why there are tens of thousands of programs available for each smartphone. Why isn't there something similar on PCs if these reflect useful features?! The overabundance may simply reflect the deficits of a WIMP interface on a small device. Many of these applications are in effect scripts that shortcut the otherwise annoying navigation of features that should be built into the device. On a PC, wouldn't most of these objectives be served by navigating to a Web page with, for example, instructions on making espresso beverages at home (a featured app in the Apple iPhone store as this is written).

Perhaps adding programs that are really shortcuts suggests a problem. Isn't an overburdened WIMP interface on a small device a bit like a poorly designed, frustrating series of touch-tone menus? Nevertheless, the advantage of a WIMP interface is familiarity; we can usually find what we want after a bit of effort. Because we've all invested so much in becoming familiar with it (like the QWERTY keyboard), it won't go away. This isn't necessarily bad; a fallback non-speech option is a requirement in most cases.

A natural evolution

A Voice User Interface should be viewed in the context of the evolution of user interfaces. It can be part of addressing the growing complexity of our use of the Web and PC applications, as well as multi-function mobile devices and entertainment systems. Perhaps the term Voice User Interface (VUI) is misleading; the appropriate approach is to make voice a complement to other modalities available, not a complete replacement. The GUI didn't drop the keyboard as an option when it added a pointing device; and Web and PC search models didn't replace the GUI as an interface.

Viewing the VUI as an evolution that enhances the prior generation of user interface innovations, rather than replacing them, is a useful approach. Certainly, when a hands- and eyes-free interaction is desired with an otherwise GUI- or text-oriented interface—for safety or other reasons—pure voice interaction provides an option. But even in this case, information can be delivered as text for later use. The issue is what best serves the user.

Speech as a shortcut

But speech technology has a further advantage. It can be like a command-line or search model, avoiding the modality and navigation that are increasingly complicating WIMP interfaces. With the objective of avoiding the need for a manual, the user interface can have a simple model—"just say what you want."

The "just say what you want" model is dangerous if there are no constraints or personalization associated with that freedom. Speech recognition has gotten continually better, but "speech understanding" that requires aspects of human intelligence ("artificial intelligence") is still in its early stages[1]. However, applications can learn about a specific user's

---

[1] The problem is perhaps one of degree. Trying to achieve general intelligence or full natural language understanding is a high hurdle. Even humans are not born with full "intelligence." We require years of training to acquire "natural language." Much of human intelligence is adaptively acquired, learned over many years of living and specifically related to what humans do because they are human. Our language and experience is closely tied to that bodily experience (see, e.g., *From Molecule to Metaphor: A Neural Theory of Language*, Jerome A. Feldman, Bradford Books 2008). A computer can't directly experience this; it has to be "told" about the human experience somehow to fully understand human interactions and language. Do we need our computers to have lips and ears, or will speakers and microphones do? Do we want computers to get angry at us when we can't master a "simple" software application? Do we want them to take years to learn our language, as humans do?

language and tendencies, making them more intelligent. Personalization is particularly natural in mobile applications where the "personal telephone" (mobile phone) is specific to its owner. Most importantly, an application can be smarter if it is operating in a known context, e.g., launching specific applications that are on a phone or PC. We will examine more deeply what the form of a consistent speech interface should be in a later section of this chapter.

The challenge is to help users get beyond initial lack of familiarity with what can be done with a VUI, and to give the interface time to adapt to the user and the user to the interface. I noticed that a voice search option for mobile phones accurately recognized "testing one two three" when first downloaded, an example of the types of things an early user might say to see if the technology works. This could be the result of tuning the system or brilliant marketing, but it is certainly the type of confidence-building effort that the industry should undertake.

The VUI is a natural extension of the history of user interface evolution. It is not an end in itself, but part of a continuing effort to make machines better serve human needs.

The speech peripheral: the microphone

It would be remiss not to note that speech technology can be defeated by poor or noisy audio input. Speech recognition can't work without a microphone. The quality and type of microphone clearly impacts the accuracy of the speech recognition. On PCs, when users do buy a microphone, they buy it for specific purposes. It may not be of particular quality if it is bought for an over-the-Internet telephone service, since high fidelity is not a feature of the conventional telephone system.

The effectiveness of a PC microphone can also depend on type. A head-mounted, close-talking microphone will provide the most consistent and noise-resistant option. A desktop or monitor-mounted microphone may suffice if it is sufficiently directional, but is more of a challenge.

The microphones on telephones are designed for human-to-human communication, not human-to-machine. When held close to the mouth, they have some intrinsic noise-resistant qualities; when used as a speakerphone, they can be a challenge for speech recognition systems. Microphones designed to be noise cancelling or highly directional can

---

Computers have the most productive effect when they complement us, not try to do what we already do well.

improve speech recognition results. One can hope that, if speech becomes an important part of the user interface, microphone quality will become part of the design criteria for mobile phones.

On a universal speech interface baseline design

Let's return to the need for consistency of speech interfaces across applications and platforms. As noted, consistency has been key in driving the acceptance of GUI interfaces; pointing and menu selection, for example, is a familiar process despite many different details. Today, that consistency is lacking for voice interfaces. It is one experience to call to get directory assistance, for example, and quite another to call a contact center and be presented with a menu, and yet another to dictate a text message.

At the time of this writing, when the average person is asked about their interaction with a voice interface, they mention call centers. To the degree there is uniformity in call center speech interactions, it is a "directed-dialog" model, where the caller is told what they can say at every step. While this is a style of interaction that a customer might come to expect, it differs with each company in its content and style. There isn't much that is intuitive about most of these interactions.

Can we establish and build on a baseline to make the Voice User Interface in applications as diverse as mobile phones and call centers as familiar and acceptable as today's GUI? What could that baseline be?

A speech-recognition baseline should:

- Be intuitive so that no user manual is required;
- Translate from one platform to another, so that one can move to a platform not used before and have a basic understanding of how to use the speech interface;
- Form the basis for understanding extensions of that baseline that may involve variations on the speech interaction; and
- Take advantage of other modalities as fallback when speech isn't an option, ideally maintaining the same mental model as the speech interface.

This chapter suggested that one possible user mental model for a mobile phone interface is "say what you want." The alternative for maintaining the same mental model when one can't talk is "type what you would say." Current implementations of this model as this is written are network-based, using the data channel, with a small client application on the device that brings up a text box when a button on the phone is pressed. The simple model is "say or type what you want into the text box, correct it if necessary, and have it take the next step that is suggested by what you entered into the text box." (A push-to-talk button can be considered part of the mental model.) Under the surface, extensive technology goes into understanding and executing the text command or dictation, much like the complexity in responding to entries in a standard text search box is hidden behind the scenes.

This approach perhaps appears to require speech recognition technology that is overly difficult for a baseline. On the surface, it would appear to require deep natural language understanding, which I believe to be too high a hurdle for today's technology. If the command were truly unconstrained, then perhaps "Say what you what" is too ambitious.

However, I don't believe a mobile phone user's request will be unconstrained. The implicit instruction is to say what they want the mobile phone or a mobile service to deliver. One doesn't walk into a pizza parlor and say to the clerk taking the order, "Is my prescription ready?" Similarly, the implicit constraint is what a mobile phone or service can do. Further, a system interpreting the statement can take advantage of the personal nature of the mobile phone to have context on the user, among other things, where they are, who their contacts are, and what they have said before.

Further, "Say What You Want" (SWYW) has a built-in constraint. It is implicitly a command. One wouldn't start dictating an email or text message in response to "say what you want," but would more likely say "dictate a message" or "send an email to Joe" first. At least, a user could quickly learn to provide some context for a command when necessary, as long as the specific way that the context was provided was flexible and intuitive.

There is another reason to believe that this isn't too high a hurdle. The system, like a pizza clerk, knows the limits of what can be delivered. If you say "pepperoni," the clerk will understand "pepperoni pizza" and ask "what size?" If you say, "size ten," you will get a stare of incomprehension. Humans use context to understand, and machines must do so.

In a VUI, there will be categories of functionality such as navigating to an application on the phone, connecting to a network-based service, dialing a contact, conducting a web search, dictating a text message, etc. (Note that dictating a message is largely unconstrained speech once one is in that mode, but dictating a message to be converted to text is not a user interface to a computer; the text is intended for a human to read, and to interpret despite errors.)

The user can learn quickly that keywords such as "search," "dial," or "dictate a message" will make the result more reliable, and the system's job in interpreting at least the general context of the message is limited to the domains it can handle. It can do the equivalent of saying "I don't understand" (e.g., a beep) if it can't categorize the request into one of these domains. Such feedback will help the user learn what works.

The trap of modality

The response to a SWYW command can be to drop the user into another speech recognition program. For example, the "dictate a message" command would presumably put the user into a specific dictation mode tuned to create an accurate word-for-word transcription. The same might be true for other commands such as music search that use specific contexts.

If a SWYW command is "call directory assistance" or another voice-channel application, one is then within the voice user interface of that application, which is likely to include more dialog. This isn't a particular problem since the user may view the interaction as a continuation of the automated voice interaction; however, it isn't clear to the user how to get back to the SWYW interface from a voice phone call. One directory assistance service addresses this modality within the application by always understanding "start over" as a return to the main menu. If a user says "phone Jane Doe" during a directory assistance call, expecting to get the usual SWYW response, the directory assistance program is likely to simply ask for a repetition. In this case, if one is done with the voice channel application, the button press that activates the SWYW application could terminate the call. "Press to talk to the system" is a simple extension of the "press to talk" part of the basic SWYW paradigm. (The issue of the voice versus data channel on mobile phones is further discussed in another section of this chapter.)

In another case of mode switching, suppose the user is on a standard call with another person and wants to conference in another person or get movie times (that is, change modes) without losing the call, how can they do so? A uniform way of doing so would be ideal, as opposed to a differing approach with every vendor or service. The button press that causes the SWYW application to listen could automatically bring it back up when on a call, but client software or the phones' operating system would have to maintain the call and provide a way to go back to it after executing the command.

Another possibility during a voice call is a "hot word." In the late 1990s, Wildfire Communications used the hot word "Wildfire" to wake up the "Wildfire Assistant" during a phone call, and today Ditech offers an always-listening hot-word technology that uses "Tok-Tok" (pronounced "talk-talk") to wake up a general command grammar. A user could be taught to preface a command with the hot word when in another mode. This extends the SWYW model, and requires that the user become aware of the option. Further, it isn't clear how the use of a hot word in a voice-channel application could drop the user back into a data-channel-based SWYW service. But appropriate innovation and cooperation between vendors—or the participation of a standards organization—may make this possible.

In another practical hurdle, the current SWYW model on mobile phones doesn't support dialog, typically dropping into a GUI after the voice request. Since network-based speech recognition is required to handle this task well today, there is enough latency that any dialog would probably have to be short. As a baseline, however, it can become a starting point for more extensive dialog as technology evolves to remove barriers.

The voice versus the data channel on mobile phones

One complication in creating a consistent VUI experience is that mobile phones have two distinct ways of connecting with computers or people. One is the conventional voice channel, and the other the data channel. The data channel supports multimodality more easily, since it can display, for example, a list of options in response to a voice request. The voice channel can deliver some information as text by email or text message if properly set up, but this is hardly interactive. As previously noted, switching from a voice interface on one channel to a voice interface on another requires some care if a consistent experience is to be maintained.

Because of the success of some smartphones that make using the data channel easier, there sometimes seems to be an assumption among market analysts that the data channel will eventually dominate the voice channel, perhaps with VoIP telephony handing voice calls. There certainly will be increasing use of the data channel, but the importance of the voice channel shouldn't be underestimated. In developing countries and even developed countries, cheap voice-only (and texting) phones may dominate market share for a long time to come. Thus, a "pure" Voice-Only User Interface (a VOUI?) may be a very important option.

Automated directory assistance services over the voice channel can be reached from any phone, and are becoming widely used. Some already offer, on the same call, weather, driving directions, stock quotes, movie times, jokes, and remember your home address if you provide it. When one calls a contact center, it is a voice call, perhaps made through a directory assistance service; the caller may expect a continuation of the same experience, a "voice site" as part of a "voice web." The SWYW model may be expected at the contact center, and can be supported by today's technology (and is so in some contact centers). The subject of voice versus data channel deserves a deeper discussion, but goes beyond the focus of this chapter.

Migration to other platforms

If "say what you want" becomes the baseline, can it translate from general telephony into other environments, such as PCs or enterprise telephone systems? The GUI has migrated from PC operating systems to Web browsers to mobile phones and more.

Consider the PC. It has a large keyboard and screen, and speech recognition hasn't made significant inroads into the general user interface, even though some speech recognition and downloadable applications have been available for years, even as part of the operating system. (The Apple Macintosh operating system has "speakable items"; Microsoft Windows has free speech recognition and text-to-speech if you turn the function on.) Yet, navigation and finding functions within some complex applications (which is most applications today) is becoming increasingly difficult and frustrating. Today's PC badly needs the "say or type what you want" model. Using it on a mobile phone could lead to a demand for the same service on PCs. And PCs have the processing power to support on-device speech recognition that allows for less latency and a dialog extension.

In enterprise telephone systems, the "how can I help you?" call routing systems are essentially a "say what you want" model. They have, when executed well (and tuned), worked for both callers and the companies deploying them. The more of these systems there are, the more individuals will understand and expect this baseline capability.

In an enterprise's internal telephone system, we are seeing the beginnings of a "say what you want" model in reaching an individual by name or function (a "speech attendant"). For example, Microsoft Exchange 2010, part of Microsoft's unified communications family, includes speech recognition for converting voicemail to text, a voice-driven auto attendant, and Outlook voice control. Speech technology could in general be extended to other functions, such as setting up a conference call. Ask for what you think the telephone system should be capable of and it should do it—"say what you want."

Other speech technologies

This chapter has focused on speech recognition as the gating technology for the growth of speech into a standard part of the user interface. Other technologies, such as text-to-speech and speaker verification, will play a role. For example, speaker verification could be used to automatically refuse certain commands (e.g., authorizing a payment) if the voice were not that of the device owner. Text-to-speech could be the other half of a flexible dialog.

Summary

The user interface for computing systems has evolved, with the Graphical User Interface a prime driver of PC and Web applications. To a large degree, it is being transferred to communications in the form of smartphones. However, on small devices, it does not transfer well, although it provides a serviceable alternative when speech can't be used.

Speech is an obvious extension of the user interface for mobile phones. The hurdle in the past has been the performance and cost of the technology; both barriers have been hurdled and continue to diminish. Services and applications using the voice and data channel will become increasingly important. They will lead to broader acceptance of speech interfaces in other communications environments, e.g., call centers and unified communications, and can transfer to other markets, such as consumer devices (remote controls for entertainment systems, etc.) and PCs.

One aspect of the acceptance of the GUI is that the industry adapted the basic WIMP paradigm that was recognizable over a range of minor variations and platforms, a "baseline" GUI design. A workable baseline speech interface should be sufficiently general that it can be learned without a manual. The typical customer service line, with its directed-dialog, menu style varies widely with each company and doesn't qualify as a baseline interface.

"Say what you want" (SWYW), on the other hand, is one possible baseline on mobile phones. Adding the ability to enter the SWYW command mode could be initiated at any time by saying a hot word, making the VUI always available, even during a call. The practicality of this approach arises from the limited context of what commands a user would expect a mobile phone or called service to "understand." A mobile phone SWYW interface with a text box wouldn't have the full flexibility of a dialog that could clarify requests at first, but it could be the baseline upon which a more flexible interface is built. Once familiar in telephone applications, it could move to other environments such as the PC.

Trends driving the adoption of speech in mobile and telephone applications

William Meisel

The previous chapter argued that good design and the best speech technology could provide a general mental model of how to interact with a device, with an implicit assumption that that interaction would first be driven by a need for a better interface to mobile phones (because of their size and the ubiquity of a microphone). The telephone in general is likely to lead wide adoption of voice interfaces if for no other reason than it is first a voice device and always includes a microphone. This section discusses trends that motivate use of speech technology in the telephone and mobile environment.

The reality is that those products and services that are financial successes in the long run—those accepted by users and the marketplace—are those that are deemed "good" designs by most observers. And, of course, it makes sense that if something is really "useful," it will succeed in the marketplace. The intent of this section, however, isn't to fully analyze market trends or the prospects for specific applications and services, but to understand how general trends in telephony impact user interface designs and the adoption of speech technology.

The need for a hands-free option

"Distracted driving" has attracted the attention of lawmakers and regulatory agencies. The issue is in part the misuse of mobile phones while driving, e.g., dialing or texting, although some studies claim that even talking on a mobile phone is distracting. It is unlikely that the use of mobile phones while driving can be successfully forbidden, since hands-free use can be essentially impossible to detect from outside the vehicle. Further, lawmakers would logically have to outlaw talking to passengers, since it's likely that that is equally distracting. Thus, using speech recognition to allow hands-free control of communications devices is an important option for mobile phone makers and automobile manufacturers to offer. Control of music systems, navigation systems, and the increasing number of electronic options within vehicles also motivates a speech interface, both for hands-free use and to avoid confusion with multiple buttons and knobs. Beyond driving, hands-free use can be convenient in other situations. It can ease the use of a mobile phone while walking, for example.

The availability of a non-speech option

Although it may seem contradictory, the availability of modes of user interaction to accomplish a task without using speech can make a speech interface more acceptable. One can't always talk in every environment, so the ability to accomplish a task, even if less efficiently, encourages the incorporation in the device and the network of many applications and features. The complexity of dealing with this expanding range of features motivates the development of a speech interface that can unify those applications and make them all seem like a single digital assistant capable of doing many things.

In addition, some information can best be delivered by means other than speech, even if it is retrieved by speech commands. On mobile phones, displaying options, text, or graphics (such as maps) is an option. Even on a voice-channel call, the ability to deliver some information as email or a text message increases the usability of the speech interface.

Making voice messages more flexible

Voice mail is a necessity for telephone calls, but it is less convenient than text messages or email, which can be reviewed at leisure, dealt with out of order, and can be easily stored and in some cases easily searched. "Visual voicemail" is a growing application, allowing message headings to be displayed as a list and listened to out of order.

Converting voicemail to text makes visual voicemail considerably more useful. Since voice mail has been around a long time, why are we seeing voicemail-to-text services proliferate now? In part, it's because handling many and long voicemails is more difficult on mobile phones than desktop phones, where it is typically easier to take notes on the voice messages.

A desire to provide Web services on mobile phones

The Web has created many successful businesses, and companies want to replicate that success on an increasingly important platform—the wireless phone. The data channel makes this possible, but in many cases, such as while driving, it is difficult to use Web services without a speech option. Further, many Web sites have not been adapted to mobile phones, and are difficult to navigate on small screens. Speech interaction may be one way to deliver services equivalent to what a visual web site delivers.

The desire to deliver Web services on mobile phones was emphasized in a keynote address at a conference in 2009 delivered by Marc Davis, Chief

Scientist, Yahoo! Connected Life: "Speaking to the Web, the World, and Each Other: The Future of Voice and the Mobile Internet." He noted that mobile is a unique medium with tremendous opportunities in terms of scale, technological capabilities, and how it integrates into people's daily lives. The mobile search use case is different from how consumers use search on the PC, and speech is a natural input method for search on mobile devices. Davis said that voice-enabled mobile Internet services will enable people to interact with the Web, the world, and each other and will change the role of voice as a medium for search, navigation, and communication. Davis emphasized the role of use context—where, when, who, and what—to make intelligent interpretations of a user query.

Participation in the same conference by representatives from other companies closely associated with the Web, such as Microsoft and Google, was an indication of the adoption of speech technology from other major Internet players. For example, speakers from Tellme, a Microsoft subsidiary, discussed what people do with voice search services and "Voice Search for Everyday Life."

The personal telephone

We have computers, and we have *personal* computers. We have telephones, and we have *personal* telephones. They are, of course, mobile phones. Unlike telephones in homes and businesses, wireless phones are almost always associated with one individual. And, unlike those tethered devices, it is almost always *with* that individual.

These simple facts are not a simple development. The personal phone is a fundamental paradigm shift. There are a number of components to this shift that can make a voice interface more effective:

1. *The user interface is space-limited:* The industry has been trying industriously to port user interface methods from larger platforms such as PCs, and has done a number of clever things to compensate for the small size of these devices. They have achieved limited success. The ultimate failure of the PC model is perhaps suggested by the plethora of "applications" for smartphones, tens of thousands in some cases. Many of those applications simply compensate for the limitations of the operating systems on those small devices and create complexity of navigation between applications by their very number. Speech recognition doesn't take up space on the screen.

2. *Personalization:* A wireless phone identifies itself and implicitly identifies you when it places a call. If a caller elects to use a service that employs personalization, the service can remember preferences

and tendencies from call to call. When a device has GPS capability, it can also indicate where you are; and localization is a powerful tool, particularly for advertising. Speech recognition can be more accurate when one can bias it toward previous choices of the individual and even their specific voice and pronunciation. Dialogs can be more compact if the user has indicated preferences by specific acts. Information from the device such as location can avoid the need to speak that information.

3. *Availability:* The device is always with its owner, making services and features always available. This will increase dependence on the device; for example, few people memorize telephone numbers any longer, since they are in their mobile phone contact list. This motivates becoming familiar with useful services, and can make the device central to the owner's activities. The device is a constant companion, and a voice interface can humanize that companion and create the mental model of a friendly personal assistant.

4. *Retention and access to information:* Because the device is always available, you may also want to be able to do things with it that you would otherwise do on a PC. You may want key information that you normally access by PC available to you on the go. That drives features such as access to email and the maintenance of a digital contact list. Accessing a list of information, such as contacts or songs, is a perfect fit for speech recognition. No "menu" is necessary and the user knows what to say. In such applications, the list is available as text and can be automatically converted into a speech grammar without effort by the user.

5. *Multifunctionality:* The portable nature of the device motivates other functions that the owner would like available when mobile. A camera, music player, navigation device—why carry multiple devices if one can do it all? The large number of options makes a voice interface for finding features increasingly attractive.

6. *Lack of uniformity:* Personal computers evolved such that one or two operating systems and suites of applications quickly came to dominate. One can usually sit down at any PC and use basic functions. Each wireless phone—and often each wireless provider—offers a significantly different experience. A voice interface can introduce an intuitive, consistent option across many devices. When the speech technology is in the network and uses the voice channel, rather than the data channel, it works for the cheapest phone (and with no data latency).

These trends demand a voice-interactive "personal assistant" model in the long run. Perhaps we should resurrect the concept of the Personal Digital Assistant (PDA) with a new paradigm.

A paradigm shift in the economics of phone calls

As this book was being compiled, mobile service providers were providing both prepaid and conventional plans that made it economical for subscribers to effectively have unlimited minutes for voice calls. Since there are only so many minutes one can be on a voice call in a day, the service providers can be sure that the minutes are in fact limited. While service providers were concerned over high usage of data channels because of high-demand tasks such as downloading video, unlimited plans also often included unlimited data.

For consumers with unlimited plans, the cost of one more phone call is perceptually zero, and the length of calls doesn't matter. That is a paradigm shift from historical perspectives on phone calls as a costly means of communication that had to be kept short. Anyone observing a teenager using a mobile phone to talk to friends probably feels that the younger generation thinks calls are free already.

To understand how that paradigm shift may affect voice usage, consider email and Web access. They are perceived as free, although customers do pay a monthly fee for unlimited Internet access, analogous to unlimited calling plans. Isn't it likely that eventually telephone calls will be accorded the similar perception that calls are "free"?

VoIP calls use the data channel and thus are part of the data plan. If VoIP usage increases, it will be hard to continue to make a distinction between voice and data on a cost basis.

As the paradigm shift toward free or low-cost telephony develops, it could have implications for automated phone services, including those using speech technology:

- *Stay on the line, please:* Customer service lines could increasingly adopt a philosophy that, once a customer's initial reason for calling is resolved, the service should encourage continued interaction to inform the customer about other options or the company's offerings in general ("upselling" or "cross-selling" being examples). Customers could be offered outbound alerts on the availability of some upcoming product or reminders relevant to the company's offerings. The longer the call, the more motivation to automate it to avoid agent costs for other than tasks for which they are required.

21

- *Call me for fun:* Some telephone "services" could be ones that customers call for entertainment, a practice certainly common in web surfing. Some calls of this genre will be motivated by conventional advertising. (See the chapter on "Advertising and Speech Interfaces.") These services could be made unique by making them interactive, as opposed to passive listening, so that callers can call the same number often, yet have a different experience each time. Part of this "conversational marketing" could be funded from the company's advertising/marketing budget, and conventional creative talent could become involved in designing the interaction.

Open-source wireless phone platforms that support speech technology

Google's Android open-source mobile phone operating system is available under a liberal license that allows developers to use and modify the code, offered through a group Google initiated, the Open Handset Alliance. The Open Handset Alliance has at least basic speech recognition available as part of the open-source package, making it easier for independent developers to economically include a speech option in their software.

The customer is king (or queen), except when they call

The increasing power of the customer, in part because of easier product comparisons and reviews on the Web, has been well documented. Treating customers well would seem like basic marketing common sense, but the theme seems to have been rediscovered recently—at least in customer service, and particularly in call centers. There are two core ideas most relevant to customer service:

(1) It is easier and cheaper to retain a customer than obtain a new one, so treat customers well and take every opportunity to develop customer loyalty; and
(2) When a customer contacts a company, the company has the customer's full attention, which is a scarce commodity in today's world of multiple media and continuous distraction. A company should make the most of such opportunities, including contact center calls, rather than treating the call as an expensive annoyance to be dispensed with as quickly as possible.

Serving the customer better with longer calls and more services can place new demands on call centers. To compound the demands, the mobility trend we've discussed will increase the use of the telephone as a way to contact companies.

Such demand can't be met completely by agents. Even if there were no cost issue, let's not pretend that all agents are effective, consistent, or available without a wait. Serving customers better requires speech technology.

The trend toward a more customer-friendly approach represents a big opportunity not only to introduce speech technology where it wasn't before, but to upgrade the first wave of speech technology. This trend may be accelerated by the availability of hosted speech solutions that provide a risk-lowering alternative for companies.

Unified Communications

Another trend is the move toward "unified communications" (UC). The core idea of UC is to integrate all forms of communication. The many features that result (e.g., obtaining email over the phone) can drive the need for a speech recognition and text-to-speech interface to make the use of these many features practical over the telephone channel. A related trend is the deluge of messages—voice and otherwise—that individuals have to manage. This complexity—as in the mobility trend— creates the need for a unifying interface, a "communications assistant." Managing communications may be just another duty for a personal assistant application.

The dropping cost of technology

Speech technology licenses, as with most technology solutions, are likely to get cheaper as volume expands, making the speech automation discussed as part of other trends more affordable. Barriers caused by the cost of speech technology licenses should diminish. And the cost of computing power to run the speech recognition software continues to drop.

Humans in the loop

Some voicemail-to-text and voice notes services use human editors to correct speech recognition errors before sending the transcription to the end user. (This is the most common case in medical dictation, some of which is done over internal phone systems.) Using editors of course increases accuracy and can reduce costs compared to transcribing speech without pre-processing by speech recognition.

One role for using editors is that the corrections can be used to improve the accuracy of speech recognition if used to tune the speech recognition parameters. Such adaptive speech recognition has long been incorporated in some dictation systems, including medical systems.

Review and adjustment of speech recognition using people occurs in call centers as well. Automated customer service applications require tuning by review of what callers unexpectedly say that causes failure of the automated system. Dialog-design experts have long used recordings of call center conversations and similar tools to adjust the speech recognition grammars to cover those cases. Speech analytics can help find the recorded calls that reflect problem cases. Adaptation can make a system get smarter over time. In some of the more difficult speech applications, editors can initially improve system acceptance and in the long term reduce the need for that human assistance.

Summary

The intent of this section was to survey trends driving speech technology adoption in mobile and telephone markets. The discussions contained some perspective on how those needs and user interface designs might evolve.

The general message is that mobile phone and telephone applications in general are likely to be the most fertile areas since they are already voice-oriented solutions. Success in this area will create a paradigm shift in the perception of speech technology and drive increased use in other areas.

Speech interfaces outside of telephony

William Meisel

The previous chapter addressed telephone applications. This chapter surveys some of the uses of speech technology in other market areas, specifically, PCs, consumer products, and industrial applications.

PCs and the Web

Telephone and mobility applications are driving the adoption of speech technology in part because telephones all have microphones, so speech input is immediately available. The small screens and clumsy input options on mobile phones also encourage speech input and output. The same isn't true of PCs, many of which don't have microphone peripherals and all of which have relatively large screens, keyboards, and a large mouse or similar pointing option. So why would speech be used on a PC?

There are several answers, and this section discusses a few:

- Dictation applications;
- Accessibility options for those who have difficulty with the conventional input/output modes;
- Applications in which speech plays a necessary part, e.g., teaching reading; and
- Voice search as a shortcut to navigating between applications and documents and as a means of access to application features.

Dictation applications

The basic function of dictation applications on PCs is easy to describe. One simply speaks into a microphone (a head-mounted, close-talking microphone is best) and the text appears as one speaks. Most software lets one dictate into existing word processing programs, so text can be edited by any means. There are many other features of these programs, including commands such as "strike that" to erase the transcription of the last utterance because there is an error or you want to say it differently. In some software, one can also say "select *phrase*" to highlight a phrase in the text, then say something different to replace it, a common editing function.

These programs can be quite accurate, particularly if one has the patience to go through the set-up stages. The software can look at the documents on your computer to pick up any specialized words and even note the way you use a word (by the context of other words around it). If you read a few passages, the software tunes to the sound of your voice, even an accent you might have. Given a chance, today's speech-to-text software is remarkably accurate.

You may have read some reviews of such software that emphasized the inaccuracy of speech-to-text software. I recently saw a video review of a speech recognition dictation product in which one reviewer with a heavy accent recited a Kipling poem to the computer while his sidekick commented in the background while standing within the range of the microphone. Needless to say, the emphasis was on the predictable errors that the system made. This isn't the first time poetry was used to test a business dictation product—nursery rhymes were used in another review.

A press release about a new version of Nuance Communications' Dragon NaturallySpeaking speech-to-text software noted that a fluent user can achieve significant productivity benefits:

> "Nuance found that the average typing speed [using a keyboard] was 35 words per minute (WPM), down from previous estimates of 40 WPM. Even more significant, the data showed that the average accuracy was only 58 percent. Dragon's near-perfect accuracy means less time spent pecking away at the keyboard and more time actually getting things done – at speeds of up to 160 WPM."

The major barrier to using a dictation program for most people isn't accuracy—it is that we are used to creating text by writing or typing, not talking. Talking is for communicating with other people, and conversational dialog is quite different than written material. Creating final written material by voice is an acquired skill. Composing that material is often a process of trial and error, with significant amounts of editing—and editing by voice is another acquired skill. Many people get discouraged before they have traversed the learning curve that makes dictation effective at creating reports.

Medical doctors and, to a lesser degree, lawyers have long used dictation, often transcribed by assistants or transcription services. They have become proficient in saying what they want to appear as text without much hesitation. A typical PC user may find that they have become comfortable with the keyboard, and, unless motivated by Repetitive Stress Injury (RSI) or a similar injury or disability, may not wish to learn to speak as they would write.

Let's step back and think about the real hurdle to creating a report or other written material—content! If we know what we want to convey, typing can work well. If we are struggling for the key ideas and how to present them, having a keyboard handy is of minimal benefit. The keyboard may even be an impediment, in that, if we begin typing, we start "word-smithing," and lose our train of thought. Ideally, we should get down all our ideas in rough form and then begin editing and organizing them.

This is more easily done if we keep our hands off the keyboard, and perhaps don't even look at the screen. Speech-to-text software can be represented, not as an alternative to typing, but as a totally different product category, "idea capture," a sort of "thought-to-text." The idea is to enhance creativity by avoiding a too-structured way of creating text. Get your thoughts or arguments down before you refine them.

The rough "idea draft" can be edited or re-spoken into the computer in a more formal style, even in a different order, while looking at the draft. This is often more efficient than trying to edit the draft. The creative process is more effective if one doesn't worry about exact wording or formatting in the initial thought process. There are text-based features that attempt to aid the creative process, e.g., outlining features, but they still can trap one in over-organizing too early.

> **Speech-to-text software can be represented, not as an alternative to typing, but as a totally different product category, "idea capture," a sort of "thought-to-text."**

Another similar use, perhaps more familiar, is "voice notes." Voice notes can also be a form of idea capture, but there is an important difference—voice notes most often are "information capture," e.g., a task for a to-do list or an appointment to remember or information that one wants to review later, rather than an attempt to formulate a creative idea. Voice notes are usually shorter and have a specific purpose. Without speech recognition, voice notes tend to get lost, because of the effort in reviewing and transcribing them manually. Some companies are already pursuing voice notes and reminders transcribed with speech recognition (or agents) as a service targeted at use over mobile phones, but the voice notes function could be part of "idea capture" software.

I am not aware of a company either marketing the speech recognition software on PCs as idea-capture software or tuning features to target it at this use.

Assistive technology

Speech technology on PCs plays an important role in providing access to PCs, the Web, and text-based material for those who are blind or visually impaired. Text-to-speech in particular can speak what is on the screen and, with a scanner and OCR software, read printed material as audio. Some software also uses speech recognition to navigate applications, web sites, and provide other aid.

For those with physical disabilities, speech recognition can be a substitute for the keyboard and mouse. Specialized software has been developed to address these needs, and dictation programs serve part of this objective.

Language and reading training

Learning to speak a new language requires hearing it. Some language training uses only recorded speech and doesn't "listen" to the student trying to learn the language, using no speech technology. But many packages do use speech recognition to score the pronunciation of a student, providing useful feedback.

Teaching young students to read requires speech almost by definition, since they can't read instructions. Some reading education software checks students' progress by such means as asking them to click on the word that is spoken. Other packages use speech recognition technology and ask the student to speak text on the screen, correcting them (or at least measuring their progress to adjust the training).

The technology demands of language training are different than applications such as speech-to-text or voice control. In the latter, the technology tries to ignore mispronunciations and come up with the closest word match. In language training, the technology is closer to speaker verification than speech recognition in that the technology tries to match what is said to how the spoken word should sound. In addition, children's voices are different from adult voices. Thus, language and reading training is a specialized area of speech technology.

Telephone style interaction on the PC

The GUI and keyboard have become so familiar that any major change in the way PCs are used faces major hurdles. However, as mobile users become comfortable with speech applications, they may ask (or a vendor may suggest to them) that they would like the "say what you want and get it" model on their PC. Thus, in the long run, speech may be a significant part of the general user interface on a PC, driven by its acceptance in other markets. Perhaps the demand will arise from a desire to communicate with a mobile "personal assistant" while at a PC.

Speech recognition in consumer electronics

Speech recognition in consumer electronics, other than mobile phones and automotive systems, seems at times to focus more on novelty than utility. Most of the devices use embedded speech technology, that is, speech technology that runs on the device itself. The amount of processing that can be done on the device (and the current state of battery technology) limits the speech technology more than when in-network processing can be done, as it is with some complex telephone applications. A short selection of some consumer products using speech recognition and text-to-speech follows to illustrate the genre, with comments after the examples. The examples were chosen to exemplify diversity, not as recommended choices, and, since products may change during the life of a book such as this, company names are not used.

Navigation Systems

While navigation systems perhaps fit into the automotive category, one where speech recognition has come to be well represented, such devices can be bought independent of the automobile. One navigation system introduced in 2009 allows consumers to simultaneously access mapping and route guidance information; enjoy various forms of audio and video entertainment; control an Apple iPod, iPhone, or a Bluetooth-enabled cell phone with natural voice commands. One can enter an address for routing by saying the city name, street name, and address number. A Text-to-Speech engine and a phonetic database enable the system to speak street names.

Music storage system

One company offered a music storage system in 2009 that supports song selection by voice commands. The product is a standalone system that supports 250 GB of tunes in the main device at a suggested retail price of $3,990. Speaker-independent voice commands and a handheld controller allow choice of songs. The software supports voice commands such as "Play *artist*," "Display by album," "Play Genre *genre*," "Queue *artist*," and allows direct selection of a specific track by saying the song name.

Travel alarm clock

One company offered a voice-controlled alarm clock in 2009. With a simple touch of a button the $24.99 product allowed for control by voice commands. One button enabled nine voice commands, such as setting the local time, alarm, and asking for the temperature.

Devices for the blind and visually impaired

A $349 reading device looks like a phone, but it's actually an MP3 and audiobook player designed for the blind and visually impaired. The product can also record voice notes and can directly convert any text file into speech using TTS.

Another reading device has both Text-to-Speech (TTS) and Optical Character Recognition (OCR) features that allow it to transform printed words into audio output, which can then be saved in MP3 or WAV formats compatible with most portable audio players. The product is priced at $699, bundled with software that provides high-quality voices.

Grocery list system

Another company offered a Grocery List Organizer in 2009, a $99.95 product that allows creating a grocery shopping list by voice. The shopping list manager matches a spoken item with one of the 2,500 food, beverage, household, beauty, and office products in its database. The device recognizes words as specific as "swordfish," "emery boards," and "lawn bags," and identifies errands, such as going to the bank, library, or veterinarian. A user can create and manage two different lists simultaneously (groceries and household, or for different people) and add products to the database (up to a maximum of 5,000 items). The battery-powered device (5" x 3 3/4" x 1 1/4" and 9 oz.) attaches magnetically to a refrigerator or mounts to a wall with an included kit.

Video games

Some video games incorporate speech technology. One video game to be published in 2009 uses speech technology to detect phonemes and maps the phonemes to audio WAVE files to create natural lip-movements in the animated game characters.

A longer-term view

These devices reflect the diverse creativity of their inventors. One wonders, however, if general-purpose and/or wirelessly connected devices will make it difficult to offer such specialized products in the long run. Some of the applications are probably available in some form for increasingly powerful mobile smartphones, for example. Certainly a mobile phone can act as a voice-activated alarm clock, for example. Scanners attached to PCs can act as book readers; there are even pen-size scanners that could interface eventually to mobile phones. If the device is networked, it can use centralized databases and in-network speech recognition for the more complex tasks. The fact that consumer applications such as these exist reflects some degree of need, and perhaps a growth in speech-enabled consumer applications if they are lower-cost software on general-purpose platforms.

Industrial applications

Perhaps surprisingly, one niche area that has significant penetration of speech recognition technology is the task of picking orders in warehouses. Workers, often driving forklifts or other vehicles, go from one bin or shelf to another filling orders, wearing headsets and devices that connect to wireless networks and are driven by Warehouse Management Systems (WMS, specialized database software). Workers give vocal commands requesting the next bin location when they have picked up an item. They can indicate if an item is out-of-stock or when they have completed picking the item, so that the WMS can keep track of inventory. If barcode scanning is part of the system, the software can use it to verify that the right item was chosen. Hands-free use of the systems also improves safety and efficiency.

The systems save money by improving inventory management and avoiding expensive problems caused by picking and delivering the wrong item. For example, at Bell Canada, field technicians must collect parts needed for a particular job. With speech recognition, current order accuracy rates are at 99.9%, which helps the field technicians to be more productive and, in turn, lowers supply chain costs and improves customer service for Bell.

The speech recognition interface isn't difficult from a dialog point of view. There are only a few commands, and repetitive use by workers lets them learn the specific commands. The application, however, has some unique aspects relative to other applications. For example, the environment is often noisy. Workers often are non-native speakers of English. For both reasons, a number of vendors use speaker-dependent speech recognition, requiring workers to "enroll" their voices and to identify themselves when using the system; speaker-dependent systems can be less vulnerable to mis-triggering by background noises or other voices. However, at least one vendor uses classical network-based speaker-independent speech recognition for the application.

Summary

The telephone is driving increased use of speech technology for reasons addressed in the previous chapter. This chapter has outlined the use of speech technology in other areas. Current use is limited and focused on specific needs.

In the longer term, design principles and technology used in telephone voice interfaces may migrate into these markets and create broader and more consistent use of speech interfaces. Some of the specialized devices, such as in consumer electronics, may become applications on more general-purpose devices such as mobile phones.

The enterprise: Customer service, voice sites, and unified communications

William Meisel

Many of the chapters in this book address specific problems in call centers or other enterprise communications applications. This chapter is intended to provide an overview of the use of speech technology in the enterprise, how it has evolved, and trends in future development. The chapter first talks about the applicability of speech technology in contact centers (customer service), how contact centers may evolve into "voice sites," and finally how speech technology serves internal communications within organizations—what vendors today are calling "unified communications."

Contact centers

There is little dispute today that speech technology can ease customer interaction with contact centers—and save money relative to using agents while doing so. Nevertheless, the slow pace of adoption of speech technology and expansion of its use once adopted has disappointed many in the industry.

And speech technology often has disappointed the caller to a customer service number as well. The perception is often that automation is blocking the customer's access to an agent. There is even a "GetHuman" website that offers suggestions on how to get around automated systems to get to an agent quickly. On the other hand, the appeal of "getting human" has met resistance as agents are out-sourced internationally, often with clearly recognizable accents, a trend some consumers respond to negatively.

Nevertheless, agents should be reserved for the complex interactions and problem-solving at which humans excel, and not be used as speech recognition devices that listen to a caller, enter what they said into their monitor, and read the results back to the caller. Other than the expense of addressing common interactions this way, it will usually provide at least a slower experience for the caller than a well-designed speech recognition and text-to-speech system. And caller authentication by voice characteristics (speaker verification/authentication) can be much more secure (and safer) than giving personal information to an agent.

Part of the reason that speech technology hasn't been more accepted by callers to contact centers is the way that speech technology automation has evolved. The first automation, often called IVR (Interactive Voice Response) systems, used touch-tone ("please press one for…"). By its very nature, touch-tone could only offer a few options, and once those options were selected, only a few more options were offered corresponding to that first choice. Other than being a slow process, callers often didn't understand which of the options related to their problem. Often the menus were chosen by business managers that segmented their activities in ways that they understood, but weren't intuitive to customers. We've all encountered frustration at times with touch-tone solutions.

Unfortunately, as speech technology was introduced, it was often approached tentatively to do only specific tasks within the touch-tone hierarchy. For example, an airline might start by automating "comfort calls" by flyers verifying their reservation and flight times, leaving other tasks to other branches of the touch-tone menu. When speech finally covered all options, the touch-tone hierarchy was still in place, a phenomenon I've called "main-menu mentality." Callers often didn't see much difference in saying one of several options rather than pressing a key if the same options were offered.

Technical solutions are easily available to reduce main-menu mentality if enterprises are willing to revisit what may have been a significant investment. Since speech technology can allow many options to be handled at once, there is no reason to limit callers to saying one of a few options presented by a prompt. For example, repeat visitors to a site may know that their option is one or more levels down from the main menu. If they barge in and say that option, there is no technical reason that the system can't move directly to that point in the dialog if designers have taken this into account. Almost no automated system at the time of this writing allows this option, perhaps a symptom of the segmented thinking encouraged by touch-tone technology.

Another aspect easily addressed by today's technology is to avoid trapping the caller in a mode they didn't intend or to allow them to simply change their mind. Most dialog applications use "directed dialog," where the system controls the flow of the conversation, limiting what the caller can say that will be understood at a given point in the conversation. A "mixed-initiative" dialog allows the caller to change the flow, e.g.,

System: How many shares would you like?

Caller: Actually, I want another stock.

At a minimum, the system should always allow the command "start over," "main menu," or similar requests to get out of a modal dialog, an approach allowed by at least one current directory assistance and information service.

Even more flexibility is afforded by "natural-language" call routing, where the prompt simply says something like, "Please tell me what you want to do at any time during this prompt, for example, ..." The underlying technology is much less capable and ambitious than "natural language" implies. It depends on the fact that there are only a finite number of tasks that the caller can accomplish on that phone number, and simply attempts to find words or phrases within the request that determine the category of task. Once determined, a dialog can clarify any ambiguities. The speech recognition is trained using typical requests to the contact center, and thus handles product names and other language specific to the company. The only downside of this technology today is its cost in terms of both licenses and professional development support from the vendor.

Some of the hurdles that managers face in improving automated customer service relate to legacy IVR (Interactive Voice Response) hardware and software. Proprietary IVR systems are being replaced by "unified communications" (UC) solutions using standardized computer technology. The "unification" results from the same basic platforms and operating systems supporting all forms of communication, including email, calls to and from the internal phone system, and customer service calls. While certainly a good idea for the long term, both in terms of cost and flexibility, the decision to move in that direction can be an expensive upgrade. Further, because UC systems become part of the company computing infrastructure rather than standalone dedicated systems, contact centers often find that their decisions are driven in part by policies of the Information Technology group—complicating decision-making.

Another potential disruption in contact center operations is the use of "Hidden Agents." Hidden Agents (HA) never come on a call; they back up speech technology (or in some cases simulate it in its entirety) by listening to a specific caller response to a prompt, sent when the speech recognition system detects a low "score," indicating that the response doesn't match responses anticipated by designers. The HA knows what the prompt was and can enter a command that interprets the utterance and gets the customer to a desired service. If the agent has no suitable option, the call can be directed to an agent that interacts directly with the customer. There are commercially available platforms that support HAs.

In part, the issue of capital investment required to implement changes has resulted in a burst of growth for out-sourced speech automation solutions. In addition to offering a pay-as-you-go model, hosting organizations can buy speech recognition licenses and telephony platforms in quantity, maintain them as part of the pay-for-usage fees, and offer expertise in application development that most companies don't need on a full-time basis. In addition, the hosted environment allows handling seasonal spikes in calls without building internal capability that is unused much of the time.

Disappointing customers can be costly. Reichheld and Thomas in their book *The Loyalty Effect* claimed: "On average, US corporations now lose half their customers in five years." It can be expensive to replace those customers—it's cheaper to find ways to keep them. Regarded on a holistic company basis, contact centers should be a major focus and investment.

Voice sites and conversational marketing

As noted in an earlier chapter, a number of trends are making voice calls as inexpensive as visits to Web sites in a Web browser. Thus, it is natural to ask if there is an analogy—"voice sites"—which, through automated speech interaction, will attract voluntary callers and attempt to engage them or complete a commercial transaction. Certainly, there are toll-free numbers that are typically oriented toward a single task and attempt to keep the call as short as possible. A voice site would be more like a Web site; it would attempt to do as much as possible to keep the caller's attention to enhance a brand or sell a product or service. Many of these sites will be paid for out of an advertising budget, rather than an operating budget. One might call the category "conversational marketing."

For example, an NBC unit has engaged a voice services company, Call Genie, to host a telephone adjunct to their popular TV game show "Deal or No Deal" (DOND). DOND was

"Voice sites" will attract voluntary callers and attempt to engage them or complete a commercial transaction

the #1 new daily syndicated series in the U.S. in its first season, averaging more than 2.4 million viewers daily. DOND viewers can play the Deal Mania sweepstakes at no cost by calling a toll-free number for a chance to win a daily prize or cash valued at approximately $1,000. All participants in the game receive a coupon with acknowledgement of their entry and are given an opportunity to register for future offers. All data is captured for future promotions and "push" marketing.

Call Genie is one of the few companies that sees itself specializing in what this section calls voice sites. Michael Durance, the company's CEO summarized in an October 2009 interview in *Speech Strategy News* the difference between typical call center interactions and processing calls at such sites:

> "The basic change is that we treat the voice application as a speech understanding problem, picking meaning out of what the caller says, not as a dialog design problem. It sounds subtle, but it is a profound shift...This is an area where voice has a real advantage over text-based search. When a user keys in a search on Google, they expect to see matching results. Maybe the results are not what they wanted, so they key in a different search term, and it could take a few tries to get what they want – if they don't give up. With speech, it is natural to engage the caller, asking them to clarify what they are looking for. It is a different paradigm, a more natural user experience that potentially gives faster and more accurate results."

Obviously, such voice sites can be independent applications, such as the DOND line. But it may also be the case that contact centers become involved, perhaps simply as a transfer destination when a caller responds favorably to a cross-sell offer or asks a question that would normally be handled by the call center. Such interactions will increase call center traffic, requiring increased automation to handle calls efficiently and economically.

Services such as automated directory assistance may also direct an increased number of calls to a call center. How many more calls? In a survey, 88% of those surveyed said they would probably or definitely use a free 411 service more frequently than a paid 411 service. As category search and general search are more widely available and better known, and as these services tune their offerings to be more effective, the increase in call volume could be overwhelming. One free independent directory assistance service—with mostly word-of-mouth publicity—already handles more than 500,000 calls per day. And major service providers are now testing free directory assistance with consumers. Since most of those services use speech recognition, using speech recognition to handle such calls may seem natural to the caller.

Using outsourced agents to handle increased volume won't solve the problem. Answering what is essentially a marketing call with international outsourcing is likely to convey the wrong message. Technical support is one thing; a product inquiry is another.

But increased traffic is only one possibility. Upper management may regard the call center as the company voice site; after all, the call center is the group with experience handling outside calls. If so, the priorities of the call center management must change from a focus on quick disposition of the call to more of a Web-site attitude to engage the caller as long as possible. This trend is an opportunity as well as a challenge.

The attention of the consumer/customer is increasingly difficult to get and hold in today's eclectic communications and entertainment environment; once a company has a customer's undivided attention, they should make the most of it. This may require use of some of the company's advertising budget or advertising agency to create creative ways to subtly (or not so subtly) advertise the company brand and products, perhaps giving the caller a chance to opt-in for "special discounts."

A voice site can also offer information, what is sometimes provided on Web sites as frequently asked questions. The point here is not to design such an application, but to indicate its possibilities and note how different it is from most call center tasks.

Another aspect of a voice site relates to an aging population. While today the PC and Web are not strangers to most older people, the phone is still both convenient and sometimes easier to use than a PC for that generation.

The challenges for contact centers translate into challenges for vendors as well. Most IVR applications are oriented toward conventional customer service, and professional services organizations often have the same orientation. In general, existing platforms have the capability to support this changing nature of calls. Vendors have the raw technology to handle these calls, but they must support them further by raising their creativity to a new level.

Unified Communications

As previously noted, from a systems point of view, Unified Communications (UC) is basically shorthand for the eventual movement of enterprise telephone systems to standard networking and Web standards and away from proprietary telephone equipment, unifying it with the same systems that provide other communications options such as email. Another unifying aspect is new interfaces between such systems, which includes options for hearing email read over the phone using text-to-speech and having voicemail delivered as email using speech recognition. Speech recognition also plays a role in controlling the exploding number of functions that such integration offers. One survey showed that only one in three business users have successfully transferred a call, and even fewer have set up a conference call. Speech recognition is used in making calls by saying a name on a contact list or by an outside caller saying a name or function to reach an individual inside the organization. Unified communications, of course, have features not employing speech technology, such as presence detection (a feature from instant messaging) and click-to-call or drag-and-drop to call or conference. Specialized phones being developed for most UC systems are essentially small PCs to allow taking advantage of such features.

In the long run, a personal assistant model—a communications assistant—will help make all these options more usable. The user should understand intuitively what a communications assistant should be able to help with—"Set up a conference call with Fred Smith, Tom Jones, and Jane Doe." The Assistant clarifies by asking questions when necessary: "Fred Smith in R&D or in marketing?" It may indicate a problem: "Jane is not available. Would you like to conference with the other two or wait? (This assumes presence detection or an out-of-office indicator.) If it's not obvious what to say, one learns the paradigm by asking: "How do I forward my calls to my wireless number?" The model, of course, is how one might interact with a personal assistant. Speech technology today can support this model in clear environments such as Unified Communications. As a bonus, one gains support for any mobile phone through the voice interface.

Recently, companies have been including the contact center in a UC model. The evolution of the "call center" to a "contact center" suggests the multiple ways customers are contacting companies, including calls and email. Thus, the unifying concept and the standardization of communication platforms applies to the contact center as well.

Mobility and the increasing use of wireless phones complicate the picture, and this is where "multimodality"—sometimes called "multichannel" customer interaction—arises. Wireless phones have screens and can display text or graphics. The phone can "remember" text messages internally, so information can be delivered directly to the phone for retention, rather than just speaking it. Must the call center take advantage of the many modes of interaction beyond voice? The option of using multiple modes of interaction during an automated or agent-supported call makes dealing with a call from a mobile phone potentially different than a call from a landline phone. Some vendors are even promoting the delivery of videos to mobile phones.

But unifying communications doesn't have to be fully comprehensive. In particular, many network-based services just attempt to solve particular communications problems. One important example is the delivery of voice mail as text, allowing it to be reviewed quickly and allowing archives of voice mail to be searched as text.

If communicating with oneself is communication, then services which allow voice notes (transcribed to text and often with categorization of the note) are another application in this category. Some services try to rise to the level of personal assistant, allowing commands such as "Create an appointment on Friday at 2 PM with Dan Smith" to create an entry in a calendar application. The utility of such applications, some of which use human agents to do the speech recognition is clear if well designed.

Summary—Enterprise trends

"Enterprise" communications applications are those that a company uses for internal communications or outside communications with customers and business associates. This chapter has provided an overview of those applications with the focus on the role speech technology can play and the form that the voice interface designs take now and how they might evolve.

Five Guiding Principles Your Callers Want You to Know

Stephen Springer

Stephen Springer, Senior Director for User Interface Design at Nuance Communications, discusses what callers want (and should be able to find) when they call a contact center. Since joining SpeechWorks in 1998, Steve has been a central contributor to state-of-the-art dialog processing. He co-designed several core DialogModules, as well as the first Name & Address collection function, SpeakFreely natural language routing, and speaker verification deployments. More recently, Steve has championed data-driven dialogs facilitating cross-media use of corporate knowledge bases. He has led the design of systems that handle millions of calls, with Transaction Completion Rates exceeding 98%.

So, you're a Voice User Interface Designer who works with corporate clientele to design automated self-service applications in Call Centers. Your business is doing well. Your customers are happy with your deployments. You are proud of your work. And you're helping to make millions of callers each year a little bit happier with their experience calling big corporations.

Then you arrive at a cocktail party, and some very nice people ask what you do for a living. Your reply is greeted, repeatedly, with cries of "*Those* things? I *hate* those things!" and "They need to hire more operators!" and "Oh, aren't they just *terrible*?" You smile and nod, and the next time you're asked, you just say, "Uh, I'm a consultant."

If this experience sounds familiar, it's because it's commonplace. While design methodologies in the Call Center industry have matured considerably over the past decade, the industry is guided by different metrics than those used by callers. Even when we track Caller Satisfaction (or "CSAT") metrics and engage in caller-driven design, a variety of influences conspire to lead us down a path very different from the one callers would want us to follow. How can we do a better job of giving our callers what they need?

The Five Principles

"The Five Principles of Quality Phone Service" aren't personal design opinions or even professional rubrics. They're what *callers* are demanding, through verbatim feedback to agents, through focus groups and usability interviews, through the very things they say to speech applications. What is perhaps most ironic about them is that they are at such a high level, and so intrinsic to the nature of Customer Service, that it's hard to imagine anyone *not* following them. These principles are:

1. Tell Me the Truth
2. Speak My Language
3. Don't Box Me In
4. Respect My Time
5. Check Your Suggestion Box.

On some reflection, you may recognize some common traps that automated systems can fall into, despite our best intentions, such that callers feel these principles aren't being followed. Let's go through them one by one.

"Tell Me the Truth" Of course no company would intentionally lie to a customer. But put yourself in the role of a consumer calling Customer Service, and hearing "Listen carefully, because our menu options have recently changed...." Do you really believe that they have? When you hear, "Your call is very important to us," do you think, "That's good to know!" or "Come on, let's get on with it..."? Even relatively innocuous messages, when placed between a caller and her goals, become obstacles. If one can't trust these opening messages, why trust later ones such as "This is a mandatory fee" or "There are no reservations available at that time"? Callers are more likely to opt out, to get "the real story" from a live agent.

"Speak My Language" Every spoken menu system essentially puts words in the caller's mouth, because we need to map a finite landscape of choices for the caller, and because speech grammars need to be programmed in advance. But when those words are lifted directly off a legal memo ("fees and exclusions") or from internal departmental jargon ("open a trouble ticket") or even from a consumer-facing website (where large blocks of text can be skimmed at a variety of speeds), they can leave behind a sour taste. Callers who feel as if they are guessing their way through a verbal maze will more likely lose patience and bail to a human being, compared with those who are encouraged to describe their situation in their own words.

"Don't Box Me In" Probably the most familiar complaint against phone systems is that callers feel they're being kept away from live agents, even though most systems actually do allow callers access. But the fundamental promise of *calling* a company on the phone is that you'll be able to speak to a person who can empathize, possibly bend the rules, and provide something that the corporate "system" (be it website, billing mechanism, or software) cannot. It's hard to overestimate the disconnect that results when a caller instead encounters another "system," one that's

deaf to emotion, and that doesn't proactively offer to get an agent on the line right away. Balancing the line between this expectation and your organization's practical need to encourage some degree of self-service is one of the trickiest acts we perform.

"Respect My Time" Call centers want to help a huge variety of callers. In an automated system, this can mean an endless succession of small updates to what was once a crisply designed system: a caveat for a new class of caller; later, a special offer; later still, a legal disclaimer; and so on. The motivation for these discrete additions often stems, not from callers' requests, but from a corporate interest, a legal concern, a governmental decree – almost any interested party *except* the caller. And these messages all take *time*. There is no "fine print" in a spoken dialog, no way to provide optional text – you either take the time to say it, or you risk not saying it. Too often, callers say, they're left waiting until they can get to the point of their call.

...an endless succession of small updates to what was once a crisply designed system...

"Check Your Suggestion Box" We measure customer satisfaction. We read "verbatim" complaints from callers as given to call center agents. We hold focus groups and brainstorming sessions and meet and debate about what will be the best caller experience. But how often do we actually ask the callers what they want, and do it? And do we let them know that? For too many callers, corporate call centers are large, faceless bureaucracies with few ways to make oneself heard.

Now the Good News

So how do we bridge the gap between that negative cocktail party gossip and all the care that we in the industry actually put into these systems? The good news is (a) we are all after the same thing, and (b) there are some simple changes we can make in our process to get there.

Are callers and corporations at odds with one another? No. Both want to do business with one another (at least, that's how they start out). Callers care markedly less about the *mechanism* than they care about just getting things done, and with a minimum of fuss. Businesses *do* care about the mechanisms, since they have different cost structures, but don't want to push one mechanism so hard that they lose a customer.

In other words, the main disconnect between businesses and callers is a mapping between specific aspects of the mechanism – the automated phone systems we deploy – and whether callers are satisfied with them. Monthly "batched" averages of ten thousand general CSAT ratings don't provide that mapping, but two small process changes will. And once we can tie CSAT ratings to specific conversations, our callers' demands will speak for themselves.

First, as an industry, we need to correlate *individual* CSAT ratings with *individual* calls. Build a tracking mechanism that generates a callback list every 15 minutes, staff a survey team to place those calls, and merge the caller's freshly reported satisfaction with the other tokens logged for that call – and then sic a few statisticians on the data to infer satisfaction drivers. Which matters more: three failed recognitions across a 20-turn call, or a lengthy database dip? We don't have to make up the answer – we can analyze what callers tell us and find correlations to events logged for their calls. By empirically validating satisfaction drivers, we can let *callers* tell us what works.

Secondly, and closely related, we can structure our deployments to allow for more minimal-pair comparisons at lower costs. Otherwise known as "Champion vs. Challenger" (CvC) architectures, the idea is the same as in any good scientific experiment: isolate a single variable (in this case, a single prompt, a single menu, etc.); control for all other possible variables; and collect metrics on overall performance so that you can ascribe different results to the changed variable. A robust, flexible CvC architecture can do away with many of the impassioned design arguments we find ourselves in, and replace them with hard evidence, supplied by the callers themselves, about what works and what doesn't. Does "this" wording do a better job of respecting the caller's time than "that" wording? Run an overnight CvC run to compare, and review your results the next day. Embracing this methodology means a fundamental re-think of most release procedures today (to keep release costs manageable), but there are surely accommodations between stability and knowledge that we can make.

Our audience is still telling us that we too frequently violate their principles. But we don't have to think of this as sour grapes. With a little ingenuity on our part, it's a terrific opportunity to revisit our callers' experience, and polish it, so that it truly shines.

When touchtone automation isn't good enough

Jeff Foley

Jeff Foley, Sr. Manager, Solutions Marketing, Nuance Enterprise Division, notes that contact center managers that think touchtone is "good enough" don't really address today's and tomorrow's reality. Throughout product launches at Dragon Systems, edocs, Atari, and Nuance, Jeff has always focused on bridging the gaps between sales, marketing, and development. Jeff holds BS and MEng degrees in Electrical Engineering and Computer Science from MIT.

Today thousands of contact centers are saving even more money by increasing call automation and they continue to keep their customers happy. What's their secret? They have designed their automation system around their customers' needs. Overwhelmingly consumers prefer an easy-to-use speech interface for phone self-service. Surprisingly, many contact center managers haven't even considered taking advantage of speech technology. Some lack understanding of just how much a great speech solution can improve their overall customer care experience as well as the bottom line. Others are nudist beach – er, that's "new to speech" – and once had a bad interaction with a speech system that misinterpreted what they were saying. Whatever the reason, most justification is ostrich-like, claiming that a touchtone system is "good enough." Good enough... for whom?

Not good enough for complex tasks

Touchtone systems may be "good enough" for tasks you're automating today, but what about what you're not automating? Touchtone automation forces you to route seemingly complex tasks, especially those that involve more than digits input—rescheduling an appointment, placing an order, contacting the helpdesk—to a live customer service representative. What's more, as speech systems prove themselves capable of handling tough tasks seamlessly, customers are choosing to use them rather than zeroing-out to wait on hold for an agent. In one Harris Interactive survey, 71% of consumers preferred the 24x7x365 availability of automated voice care.

> Touchtone systems may be "good enough" for tasks you're automating today, but what about what you're not automating?

Not good enough for

routing

Touchtone systems aren't always "good enough" for routing a caller to the right place. More often than not, touchtone systems present barriers to today's savvy and demanding consumers. Lengthy main menus asking you to "please listen carefully to the following eight choices" break the social conventions of interacting over a phone—you talk, I respond, you answer. Organizations are forced to create categories to present to callers and then callers are strained to guess the category that fits. As a result, too many callers either get overwhelmed or get lost in a maze of menu options until they finally press "0" to reach a live person. In fact, many have become trained to 'zero-out' immediately rather than wade through the options.

The latest speech-based solutions eliminate touchtone mazes by providing accurate, cost-effective routing to a caller's destination. Rather than a long list of options, callers are asked an open-ended question such as "In just a few words, tell me how I can help you today." The caller can respond with something like "Yeah, ummm, I have a question about a bill" or perhaps "I'm moving and I need to change my address." They are then sent directly to the right destination—possibly a touchtone self-service application, a speech-enabled application, or an agent who can help them with their specific request. Callers are more satisfied because they want to focus on their own needs rather than having to guess which bucket to explore to find an answer.

Not good enough for today's callers

Finally, while touchtone systems may have been "good enough" in the past, they really haven't kept up with present-day caller expectations. Consider the phones in use when touchtone systems came into being. Hold the receiver in one hand, push buttons with the other. Isn't touchtone great? Listen and push, listen and push.

Now consider modern phones. Many home phones are cordless—all the buttons are on the handset. Callers can go back and forth either pushing buttons or listening, but never both at the same time. What about the mobile phone users? In a recent Harris Interactive survey, consumers responded that they use their mobile phone to call customer service about 70% of the time. They're too happy with their iPhone – you know the one with no buttons on it. Some mobile phones have the ABC, DEF labels on the keypad, but many do not; callers with a mini-keyboard phone like a Blackberry are counting through the alphabet to spell out a name. Meanwhile, more and more cars now come with telematics capabilities. While sitting in traffic, your customers are calling your customer service number—often dialing by voice. If they're pushing buttons on their dashboard, it's probably to hit zero and talk to an agent. More and more consumers are using touchtone-unfriendly devices.

Good enough is not good enough

The ability to interact with your customers in a conversational manner, the way they naturally speak, is a powerful driver for increasing customer satisfaction. Rather than being forced into rigid menus, customers using self-service speech solutions enjoy simplified dialogs. In addition to reporting a much better caller experience, they tend to have shorter calls and complete them within self-service, increasing automation rates and reducing costs. By enabling more conversational input, speech addresses the needs of today's consumers. In short, touchtone simply isn't good enough for today's consumers.

Making Self Service Preferred

Kevin Stone

Kevin Stone, senior director of products, Nuance Communications, discusses how to create a self-service experience automated with speech technology that callers will prefer to an agent ("preferred self-service"). Prior to joining the Nuance team in April 2007 as part of the BeVocal acquisition, he served as vice president of marketing at BeVocal where he was responsible for all of BeVocal's product management and marketing activities. Stone received a bachelor's degree in electrical engineering and computer science and a master's degree in electrical engineering from the University of California, Berkeley.

After many years and many unsatisfactory experiences with poorly implemented self-service applications and Voice User Interfaces (VUIs), most callers have come to think of a live agent as their preferred interface of choice for handling most of their needs. If implemented correctly, however, self-service can be the *preferred* method of interaction with contact centers.

What does a preferred self-service application look like? Visit your nearest airport to see a shining example of a self-service application that is preferred by most users: the airport kiosk.

Waiting in long lines for a ticket agent is one of the least pleasant experiences air travelers face at the airport. To allow passengers to avoid waiting in long lines, airline companies offer the airport kiosk—a faster, self-service alternative. Queues at kiosks are rare. The devices are intuitive, uncomplicated, and easy to learn and use—even for a first-time flyer. Login is straightforward. Simply swipe a credit card or type in the passenger's name. Only the appropriate or relevant options are presented at any given point in time, based on what the traveler is attempting to do, such as print out boarding passes, change seat assignment, or upgrade to Business Class. The kiosk gives travelers a feeling that they are in control, that they can do what they want quickly and easily. And if, for some reason, travelers require an agent, they can take the ticket information they printed out to the counter—they don't have to provide that information again. It's a seamless experience. For most travelers, the kiosk is *preferred* over waiting in line for a "live" agent.

Many of the same characteristics that make airport kiosks preferred over ticket agents can also be applied to the contact center, creating an environment in which self-service applications with effective VUIs are preferred over live agents. A basic set of requirements for making self-service preferred includes intuitiveness, relevance, core functionality, the ability to reach a live agent when it's appropriate, and the ability to treat related interactions as a single transaction or continuum.

Intuitiveness

The goal of any product design is to make it easy to use. The user interface must be simple, obvious, and consistent. The path and the process to getting the desired information or result should be obvious. Preferred self-service applications require no user manuals and few explanations—they should be intuitive to the consumer.

For a VUI to be simple and obvious, it needs to understand whatever the caller might say through a natural language interface, or it must set a proper context by instructing the caller as to what can be said. In the latter case, even if the commands seem obvious, the VUI must be flexible. For example, if the system prompts the caller to say "Account Balance," it should understand multiple variations of this response such as "I want my Account Balance" or "Last Statement Balance."

For a VUI to be consistent, common commands must be universally accessible and work the same way across all points in the self-service application. Among commands that should be consistent are Operator, Go Back, Repeat, Go to Main Menu, Start Over, and Help.

VUI functionality must also be consistent. If, for example, a mobile phone provider allows its subscribers (through a self-service application) to add, change, or delete a text messaging plan, which carries a monthly reoccurring charge (MRC), then other MRC plans should also be able to be added, changed, or deleted—not just added.

Relevance

Preferred self-service is about leveraging what you know about your customers to better serve them. Care information, as well as offers presented, should all be relevant to the individual customer. VUI relevance can be reactive or proactive.

In an example of reactive relevance, a customer who recently purchased an item from a company calls that company's main sales number to follow up on the order. Based on Automatic Number Identification (ANI), the IVR system performs a database look-up and provides relevant order status details to the caller.

Oftentimes, leveraging non-voice channels, such as email and short message service (SMS), can be useful to proactively deliver relevant self-service. For example, suppose a new Voice over IP (VoIP) subscriber is awaiting the delivery of a router to activate service. The service may have been ordered through an IVR or live agent. In either case, the system can send an SMS message to the new subscriber's cell phone to confirm the order and periodically update the order status.

Core functionality

Consumers know what they want and what the outcome of their interaction with the customer care center should be. The self-service application should facilitate that outcome, not hinder it. The application should rapidly understand the caller's intent and discover what the caller wants to do. It should gather the appropriate information to achieve that outcome and then perform the necessary action via the most appropriate channel. To be preferred by end users, it must be available and provide consistent information 100 percent of the time.

For example, a common belief held by callers suggests that live agents have access to more or different information than IVR systems. If a caller indeed receives information from an IVR system and subsequently is told something else by a live agent on a follow-up call, the caller will likely prefer to talk with a live agent in the future. Similarly, if the core functionality does not keep up with changes in the business, then the self-service application will soon lose its preferred status.

Reach a live agent when it's appropriate

...a common belief held by callers suggests that live agents have access to more or different information than IVR systems...

While well-designed self-service applications can be faster and more efficient than their live-agent counterparts, some situations ultimately require interaction with a live agent. Preferred self-service has the ability to recognize when that option is appropriate and transfers the caller (and any related information) to a live agent quickly.

Appropriateness is a combination of system design and user interaction. For example, the system may collect only a portion of the information needed to file an insurance claim before transferring the caller to a live agent. Or, if the caller is having difficulty providing certain information in the claim application, the system should proactively route the caller to a live agent.

Treat related interactions as a single transaction or continuum

Consumers contact customer care centers for many reasons: from billing inquiries to order status to information about products or services. Many of these interactions, such as verifying an account balance, can be completed quickly in a single, self-service transaction lasting less than a minute. Other transactions are more complicated with multiple steps that may not be completed in a single interaction. Preferred customer care demands that the system view the entire sequence of related interactions as a single customer experience—a continuum.

For example, when booking a flight, the caller may not have all the necessary information handy. If the caller hangs up and calls back, the system should resume where the caller last left off.

Self-Service as a Preferred Service

Today, business is being conducted in what can be referred to as a "Care 2.0" environment, where companies are required to make the most of every interaction in order to satisfy the expectations of increasingly savvy, highly mobile customers who feel entitled to great service. This environmental shift necessitates a change in the way companies approach care and respond to customers. Self-service—as a preferred service—is the solution to this challenge. Consumers who have had a preferred self-service experience are likely to return to that application again and again.

The Role of Tools in Caller-Centric VUI Design

Stefan Besling

Stefan Besling, vice president of engineering, VoiceObjects (a Voxeo subsidiary), discusses tools that help customize a caller experience. As Vice President of Engineering, Stefan Besling provides leadership for VoiceObjects' software development team and is responsible for delivering product releases. Prior to joining VoiceObjects, Stefan worked in Silicon Valley for a number of years first as co-founder of a Philips Electronics spin-off focused on voice-controlled multimodal applications for mobile devices and then as Director, Voice Applications, for YY Technologies, a company providing a Web and email-based self service platform. Stefan has a background in speech recognition technology and held development and research roles at Philips Speech Processing and Philips Research.

Did you ever call a self-service number just for the fun of it?

If you are in the industry, maybe you have – but most people never would. They call a self-service number because they have a task to complete: They want to track where their package is, pay a bill, or check on flight delays. Talking to an IVR is a means to an end.

Thus the primary job of a Voice User Interface (VUI) designer is to make callers as effective as possible in completing their tasks.

Designing an application this way is a challenge – yet designers don't need to do it with their bare hands. Rather than relying purely on Excel and Visio they can utilize tools that make them more efficient. In this chapter, we want to explore some of the requirements for such tools.

Over time, a variety of concepts have proven to be useful in designing effective VUIs. A good tool ought to come with built-in support for these concepts, so that designers can focus on making callers achieve their goals instead of just making the application work.

After all, that's what a tool is there for.

Say the word

The audible surface of any voice application consists of its prompts. Significant effort is usually expended on writing and recording them – as it should be, since the caller's input is in response to these prompts. What callers do or do not say is strongly influenced by what the prompts explain or omit.

Tools that support the VUI design process should therefore provide sophisticated prompt management capabilities going way beyond the mere creation and maintenance of prompt lists, including:

- Randomized prompting that allows for a less machine-like impression should be a native concept.
- The same holds true for prompting based on how often the caller has visited a certain spot in the application. This allows for a natural, staged presentation of information.
- Multi-lingual applications should be supported, and it should be easy to add an additional language to an existing application.
- Similarly, the use of multiple alternative prompt sets, e.g., for novice versus experienced users, or for different personas, should be a built-in notion.
- Routinely, dynamic data such as money amounts, dates, etc. need to be read out. This is usually best done by concatenating a sequence of audio prompts with the appropriate intonations. The exact way of doing it depends on the type of data and the language used, so the tool should offer pre-defined mechanisms for standard cases as well as a way to extend this to custom needs.
- Finally, the number of audio resources in real-life applications can get very large. So a mechanism to validate the resources should be available that allows designers to check whether all required files are available.

To thine own self be true

VUI designers know that consistency is crucial. Confused callers are less effective and produce more recognition problems. Therefore the same commands should cause the same actions everywhere in the application.

Tools should provide an easy way to obtain such consistent behavior. Most commonly this is done by adopting an object-oriented approach. Designers can define global commands such as "Main menu" or "Representative" that are automatically available throughout the application. The same typically applies to universal navigation commands such as "back" or "repeat."

Consistency is crucial. Confused callers are less effective and produce more recognition problems.

Similarly, the same caller problem such as a No Input or No Match event should trigger the same structural behavior. A frequently used pattern is to play a generic prompt for the first occurrence ("Sorry, I didn't get that."), play a help prompt for the second occurrence ("You're in the main menu. Please select from among these options…"), and transfer to an agent for the third occurrence.

Tools should allow designers to define the desired event handling pattern globally and then only adjust or augment it individually, e.g. to include context-specific information. Not only does this ensure consistency, but also significantly reduces the amount of work that goes into event handling.

I got rhythm

Once the bases of natural prompting and consistent interaction are covered, designers can explore truly caller-centric approaches.

Did you ever notice that when you say your phone or bank account number, you have your own rhythm? Some of us will use streams of digits ("one two three four") while others use blocks of numbers ("twelve thirty-four"), and so on. This goes beyond articulation: we actually think of the numbers in this way and have a hard time recognizing these same numbers when they are presented to us in a different rhythm. The same often holds true for synonyms such as "Amex" versus "American Express."

Tools should be able to recognize not only the caller's input but also their rhythm so that the same rhythm can be re-applied, such as when confirming an account number. A subtle touch, yet maximizing effectiveness with a truly caller-centric design is the combination of many small changes that amount to a big difference.

Stay focused

In many cases, applications can identify the caller based on ANI or on an initial authentication. This opens the field for personalization and the anticipation of caller needs. Someone who reported a problem with their Internet connection yesterday may call in again today to check on progress. Mobile subscribers who just received their monthly bill may call in with a question on some of its items. By integrating with other backend systems, caller-centric applications can offer customized menus or personalized information pop-ups. In fact, the interaction strategy can range all the way from directed dialogs in which the caller is led through a series of interactions to mixed initiative in which callers are in charge of what to do, when, and where.

Tools should support standard interfaces for communicating with backend systems, as well as the means to dynamically alter the application flow and logic based on their input. They should also handle both directed and mixed initiative dialogs and leave it to the designer to select the most appropriate one for each situation.

All together now

No prompt is an island, and no interaction happens in a vacuum. To get a feeling for how a dialog unfolds, how consecutive prompts connect, and how callers are likely to respond, it is important for designers to call an application as early as possible. Changes can then be made in an iterative process that improves the design with each turn.

Tools should support this by embracing rapid prototyping, and by providing easy ways of deploying "in-progress" applications so that the designers can call, tweak, and debug them. This also includes the use of "placeholder" grammars that are easy to define yet fully workable. Usually they consist of just a few typical caller utterances – simple enough to define even for someone with limited grammar development background, yet sufficient to convey the essence of the caller interaction at this specific step.

Close the loop

Sometimes the best intentions still don't make things right. Even after lots of internal and friendly user testing, the success of a VUI design can only really be evaluated once callers actually use the application. Do the prompts encourage them to say the right things, or are there lots of No Match events? Can they successfully complete their tasks, or do they routinely fail at certain spots? Do they get what they need quickly, or do they spend a lot of time finding their way around? Which are the most frequent tasks callers need to complete, anyway?

Tools should automatically collect pertinent data to provide the relevant answers in reports highlighting "event hot spots" or dominant paths through an application. VUI designers can then take this information as the basis for another round of improvements, turning the wheel in the application's life cycle.

Summary

Callers of self-service applications have a specific task in mind that they want to complete. Designing a VUI that makes them as effective as possible in completing this task is a challenging—yet highly rewarding—job.

Traditionally many designers have worked on VUI design with the electronic equivalents of pen and paper. Today, however, tools are available that can boost designer efficiency by incorporating crucial lessons learned from real-life applications into the vocabulary of basic concepts. One such tool is the Voxeo VoiceObjects platform, which can be downloaded free at http://www.voiceobjects.com/free. It provides a Storyboard Manager to work with prompts, a Desktop to assemble, test, and deploy applications, and the Infostore/Analyzer combo to collect detailed usage data and present caller behavior reports.

In the end, there's one thing that counts: Efficient designers make for effective callers.

Voice User Interface Design: From Art *and* Science to Art *with* Science

Roberto Pieraccini and Phillip Hunter

Roberto Pieraccini, Chief Technology Officer at SpeechCycle, and Phillip Hunter, Founder and Principal at design-outloud, discuss how design alternatives in parts of a VUI dialog can be evaluated objectively based on statistics compiled during use of the application. Pieraccini has worked at CSELT, Bell Labs, AT&T Labs, and IBM T.J. Watson Research, led an R&D team at SpeechWorks, and written more than 100 papers and articles. Hunter has designed scores of applications, built successful teams at Intervoice, Voice Partners, and SpeechCycle, led Fortune 500 projects, won awards, and invented processes and tools at several previous companies.

ART HISTORY

Over the past 10 to 15 years there has been a tendency to consider the design of voice interactions and voice user interfaces (we will refer to both as VUI)

> **There has been a tendency to consider the design of voice interactions and voice user interfaces primarily an art**

primarily an art. The ability to create aesthetic and functional experiences was made possible and mediated mostly by the knowledge gained in having built dozens of applications. The evaluation of VUI solutions was based by and large on the introspective ability of the designers and their capacity to predict consequences and to make the right choices. Challenging and new situations required consulting with other designers and, more often than not, several possible solutions were discovered, not counting those proposed by the customers and by anyone else on the project with a self-determined knack for VUI.

However, this view of the art of VUI only partly parallels the applied education and craft found among practitioners of the classic arts. In those fields, science is also a tool and even a source of inspiration. For example, the learned sculptor understands how to wield force in cooperation with the characteristics and limits of steel and stone. But, along with other arts, there is a singular, common, and hard-to-know component in the final product: the emotional response produced in those that experience it. Yet, that causes the artists to know and master their tools and domain even more—so they can understand, gauge, and even manipulate the response their art provokes. So, art and science are companions in the pursuit of excellence and desired effect, complementing and even relying on each other. However, while VUI producers have long used the tools and techniques of the trade, rarely has there been a full use of science behind voice interaction and an application of what can be learned from the responses provoked by the voice system. This ignorance of the data and its meaning has been a shortcoming and even a failure point for many speech deployments.

The key to improving this situation is to first go beyond the norm in training VUI designers, ensuring that they are designing from sound principles that are founded outside the borders of the IVR world. Then, that expertise is prepared to make full use of a powerful tool in the pursuit of high-quality caller experience and application performance.

THE AGE OF SCIENCE

From its very early stages, SpeechCycle adopted approaches that better applied scientific evaluation to design: choose the design—or designs—based on sound principles that optimize well defined measurable criteria such as, for instance, automation rate and average handle time. And that is done in an objective manner by using statistical evidence on large amounts of live data. This is especially important as speech applications grow in scope and complexity. Calls to SpeechCycle applications involve dozens of turns and can follow hundreds of dialogue paths.

The idea of making design choices based on the statistical optimization of objective measureable criteria is not new. The approach of using machine learning—reinforcement learning in particular—for the optimization of dialog systems was first introduced by Esther Levin and Roberto Pieraccini in 1997 [1] while at AT&T Labs, and then extended and improved by other researchers such as, among others, Steve Young [2] at the University of Cambridge, UK, Olivier Pietquin [3] at the University of Mons, Belgium, and Oliver Lemon [4] at the University of Edinburgh. For a current discussion on the implications of reinforcement learning for commercial dialog systems see a recent article by Tim Paek (Microsoft) and Roberto Pieraccini (SpeechCycle) published in *Speech Communications* [5].

Essentially reinforcement learning uses algorithms that favor the choice of design alternatives that increase the average automated agent performance, defined as a function of the objective criterion to be optimized (e.g., average automation rate, call duration, number of no-matches, or a combination of them). While the application of reinforcement learning to the full design of dialog systems, although possible [1], is still impractical [4], SpeechCycle adopted a partial design approach where only a small but significant number of *exploration* points are set in a call-flow. Each exploration point corresponds to two or more reasonable design alternatives. Exploration is implemented by a platform feature, called "Contender" at the time.

CASE STUDY I

As a simple example to illustrate the power of statistical exploration and analysis we show the use of this model for the optimization of the portion where the reason for the call is captured within a large technical support application deployed for cable TV providers. This part of the application was underperforming and we sought an increase in the numbers of callers successfully conveying the reason for their call. Two new structures, "Alt-1" and "Alt-2", were determined to be viable candidates and both were placed into the application in parallel to the existing "Original." The Contender feature of the RPA platform randomly selected one of the three options for each incoming call using a uniform probability distribution. The selected option was marked in the logs of each call. After about 26,000 calls, the differences in the automated performance between the subsets of calls corresponding to each selected option were deemed to be statistically significant based on a standard significance test. Additionally, we examined a number of typical measures, like grammar accuracy, hang-up and opt-out rates, number of successful problem captures, and number of overall successful calls. The results clearly indicated that Alt-2 was the better approach with relative reductions in opt-outs (18%), hang-ups (23%), and improvements in speech recognition performance. So, while applied VUI principles and art were certainly needed to create the dialogue structures and prompts used for this experiment, in the end it is also the data-driven scientific method that determined which approach was best and should be used in full production.

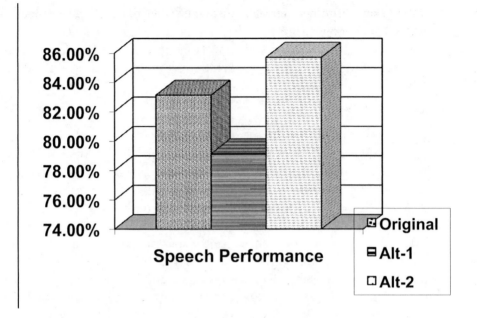

86.00%
84.00%
82.00%
80.00%
78.00%
76.00%
74.00%

Speech Performance

☒ Original
☐ Alt-1
☐ Alt-2

CASE STUDY II

Even when competing VUI designs are not being tested, SpeechCycle constantly and thoroughly analyzes large amounts of data produced by caller interactions. Data mining and analysis, mostly performed with the help of automated tools, results in useful information about caller responses that influence the evolution of the application. For instance, the analysis of out-of-grammar (OOG) responses is particularly interesting for the improvement of the system performance. Often callers give adjunct information or answer a question that has not yet been asked. As humans, we do this frequently:

Person 1:   "Where's your car parked?"

Person 2:   "I took the bus today."

Traditionally, spoken dialog systems crash and burn with those kinds of interchanges, despite them being a stalwart conversational aid between people. However, with the help of data analysis and good design practices one can revise the system and properly handle such utterances, as in the following example:

IVR: Which one can I help with? Just say "my bill", "tech support", "orders", or "appointments".

Caller: "Internet service"

IVR: Okay, Internet service. Just say "my bill", "tech support", "orders", or "appointments".

Caller: "tech support"

For the above example, significant amounts of data indicated that such utterances as "Internet service" were common even though the directed dialog prompt did not mention them as a possible choice. Design analysis indicated, of course, that the utterances were in context for the application. Thus, SpeechCycle designers created the simple yet uncommon method of handling them that is illustrated above, which allowed callers to flow through the interaction smoothly and successfully. The detection, analysis, design, and implementation of this instance took about 40 person-hours over three weeks. Without the combined approach of data collection, analysis, tools, and expert VUI design happening in a very timely fashion, such a solution would likely have occurred, if at all, several months after deployment and would have depended, in traditional tuning, on whether a speech scientist happened to notice the pattern. Traditional tuning performed sporadically offers only haphazard responses to often stale data. On the other hand, constant evaluation and analysis can lead to remarkable short-term improvements.

CONCLUSION

A better partnership of data and design and of applied principles and scientific evaluation using careful measurements and automated continuous analysis, can ensure that appropriate application evolution occurs promptly. It is this reliance on scientifically approaching performance evaluation and design decisions that drives the next evolutionary step for all speech applications. Art? Yes, but not an art without scientific discipline. The exploration at SpeechCycle has shown that scientific assessment and analysis can effectively augment and evaluate the art of VUI design.

REFERENCES

1. E. Levin and R. Pieraccini, *A stochastic model of computer-human interaction for learning dialogue strategies*, Proc. of Eurospeech 1997, Rhodes, Greece, September 1997.
2. S. Young, *Talking to Machines (Statistically Speaking),* Proc. of 2002 International Conference on Spoken Language Processing (ICSLP), Denver, Colorado, September 2002.
3. O. Pietquin, *A Framework for Unsupervised Learning of Dialogue Strategies.* Presses Universitaires de Louvain, SIMILAR Collection, 2004.

4. O. Lemon and O. Pietquin, *Machine Learning for Spoken Dialogue Systems*, Proc. of Interspeech 2007, Antwerp, Belgium, August 2007.
5. T. Paek and R. Pieraccini, *Automating spoken dialogue management design using machine learning: An industry perspective,* Speech Communication 50 (2008), pp. 716–729.

Get to Know Your Caller with Natural Language Data Collection

Michael Moore and Michelle Winston

Michael Moore, Business Design Analyst, and Michelle Winston, Sales Engineer, West Interactive, discuss a way to avoid being overly prejudiced by existing DTMF-inspired prompts. Mike is a senior member of the West's Expert Solutions Group. His responsibilities include speech application analytics, VUI design, usability analysis and testing, and designing integrated applications that incorporate speech/touch-tone and live operator support. In addition to his expertise in call automation, Mike also has eight years of call center experience and has been involved in virtually every aspect of call center management. Michelle is a sales engineer with West's Expert Solutions Group. Before joining West, she was with The Gallup Organization, serving as product manager for its metrics delivery platform. She has expertise in usability analysis, user interface design, and performance measurement—particularly in evaluating performance of customer service interfaces.

Traditional approaches to VUI design are conducted primarily from the inside-out—that is, from the designers' point of view ...out to the customer. Often that perspective is based on a limited pool of caller information. Existing business processes, agent disposition data, and logs from applications are primary sources of design cues. But do these sources alone provide an accurate depiction of caller behavior? We contend that another, highly effective, approach to VUI design is to learn customer requirements directly from the mouth of the customer. Adding a new requirements research tool to your repertoire will lead to better design, which will delight your customers and reduce costs per interaction.

Traditionally, design for new applications has largely been driven by information gathered from legacy applications. In essence, the new speech applications still have their roots in DTMF design. These are directed-dialog, menu-driven applications that guide callers through a series of prompts. The prompts are intended to help callers complete their objective—either through automated self-service functionality or via an agent. Caller behavior data from directed dialog menus have always told us what options callers select most often. Unfortunately, the data set is restricted to the options that we offer the caller. We don't get to hear why individuals are *really* calling, nor do we get to hear it *in their own words*. Reviewing caller behavior data, conducting focus groups, and administering usability tests give us some insight into what callers are thinking as they are navigating through these systems. However, since callers are operating in a controlled environment, data derived from the tests don't really tell us what callers are thinking *at the precise moment they call*. Furthermore when we observe callers in a controlled environment like a usability study, they are motivated to get through the test system in accordance with the objectives stated by the test coordinator. A restricted list of options that are indicated by a confined dataset limits visibility into caller intent. It raises the question: If we don't know what our callers want to do, how can we design self-service applications that they can use? It's no wonder there are so many people banging away on the zero key!

The emergence of natural language speech technology allows designers to rethink not only the concept of a main menu, but also their perceptions of what callers *really want to do*. This innovation in speech science presents us with a new design tool. Today, designers can expand their field of vision just by asking customers why they are calling. Callers to natural language applications respond to an open-ended question like, "How may I help you?" with a range of naturally worded responses. They have the freedom to give any answer they wish; they're not limited to the list of choices the designer gives them. By offering callers an open-ended response, callers—in their own words—reveal their true intent. Simply hearing the voice of the customer and understanding why they are calling provides a compelling source of information to drive UI design.

Copious amounts of customer utterance data are required to design and develop a natural language application. A common technique to capture data is to record an open-ended question, place it at the top of the call flow, and record the caller's response. The recordings are then transcribed, analyzed, and used to build the corpus of utterances needed by the statistical language model (SLM). But these utterances have more value than just for the SLM. Indeed, they can yield valuable information for the VUI designer about what callers want, how they make requests, and the jargon used when making requests. Ultimately, designers can use these customers' free-form responses to evaluate existing functionality for making design decisions. For instance, they may wish to grow functionality in some areas while streamlining or reducing it in others. They may alter prompt verbiage and speech grammars in order to make the application and user experience more efficient and friendly. Or the utterances can be used to provide a clearer idea about why people are calling for marketing and customer-service purposes. A rich set of customer utterance data can be used to drive design decisions for multiple application objectives.

The utterance data can also indicate ways in which callers articulate requests. Take a situation, for example, when a patient goes to a doctor. Upon conclusion of the visit, the doctor's office will file a claim with the patient's insurance company. Later, the patient may call their health insurance company to find out if their "bill was paid." However, a typical set of menu options may ask the caller if they want to "check on a claim," "find out when eligibility begins," or "indicate what benefits are covered." Unfortunately, while "claim" may be common jargon in the insurance business, it is not necessarily a term universally known to the public. The caller may become confused and want to speak with an agent. The question that the VUI designer should pose is, "When a caller wants to check on a claim, *how do they ask the question?*" Natural language data collection can provide an answer. In a recent test conducted by West Interactive, our utterance information showed us that a large percentage of callers were asking to "check on a bill" they received—not inquire about the "status of a claim." This proved to be valuable information for us in designing future applications of this type.

Callers are often able to articulate their full objective when presented with a free-form prompt. A key benefit of natural language technology is the ability to flatten menus and drop callers deep into a call flow if their answer is specific enough, thus saving time and improving caller experience. During utterance analysis for a test application, we heard customers saying, "I want to pay my bill with a credit card." With that single response, we know that the caller wants to *pay their bill*. And we also know *how* they want to pay their bill—with their credit card. So the VUI design doesn't need to ask the caller for method of payment; we already know it, thanks to the open-ended natural language prompt.

> **A key benefit of natural language technology is the ability to flatten menus and drop callers deep into a call flow...**

Another benefit natural language prompts offer is that they help VUI designers to gain visibility into the full range of requests a caller may have, and track them for analytics and future VUI design purposes. This is helpful to both the customer service and marketing departments, the latter of which is always interested in knowing the effectiveness of their advertisements and promotions. In the customer service case, information gleaned from responses to a natural language prompt can help VUI specialists to design "intelligent" applications for callers that are simple, relevant, and intuitive.

Much can be learned from the natural language data collection process, and those discoveries can be applied to improve VUI design and application analytics. What can NL data collection and analysis tell about the existing functionality of our self-service applications? While we may be confident we know the needs of our callers, how can we know for sure? By opening the range of responses to a virtually unlimited array of utterances, we gain perspective and develop a better understanding of customer objectives. By paying attention to those objectives and designing an application accordingly, speech applications can provide better service and do so more efficiently and effectively. Customers are calling to ask for something. The natural language prompt provides VUI designers with the means to hear exactly what that is.

Using AI to open up the range of services that may be viably automated using speech

Kirsty McCarthy

Kirsty McCarthy is vice president sales & marketing and Co-founder, Inference Communications, which licenses speech application development tools focused on call centers and also develops applications for customers using those tools. She describes one can use of artificial intelligence techniques to allow successful quick and cost-effective development of a call center application that was only going to be used during a four-week span. Kirsty has worked in the technology and contact area space for 20 years. She is on the board of the CCMA (Customer Contact Management Association) in Australia and was instrumental in the establishment of the Australian chapter of AVIOS.

At Inference Communications, we use advanced artificial intelligence (AI) technology to deliver value across all stages of the speech recognition application lifecycle, with the intent of delivering speech recognition services to a broadening market. Our grammatical inference technology confers a number of advantages on designers, developers, and application tuners, which are evidenced in implementations using this technology and will be discussed here in the context of an actual deployment in Australia.

Grammatical inference is a technique whereby grammar files are automatically generated as a result of the input of 'sample' utterances. It means that only a small number of sample phrases are required in order to generate sophisticated, optimally-spanning grammar files. This means that:

...only a small number of sample phrases are required in order to generate sophisticated, optimally-spanning grammar files.

- Prototype solutions may be generated with minimal effort and many optional call flows are generated to enhance the design and usability testing phases;
- The development phase is very fast; and
- Tuning is semi-automated—deployments improve the more they are used. Using grammatical inference, speed of development does not mean a compromise in quality (as exemplified later in this chapter).

This chapter will focus on an actual deployment of a solution for the Australian Football League in Australia—the nation's largest sporting body—that delivered a streamlined approach to managing customer and supporter enquiries for information during their four-week Finals Series.

The business problem was that the AFL call centre and switchboard is overrun during the Finals campaign with callers from supporters requesting information about the games, locations, tickets, TV and radio broadcast times, and information about special events. The AFL wanted an automated service to handle the call volume, but the service is high profile and an automated system must deal well with callers who were expecting to speak with a live agent. Commercially, the payback for the service—largely a reduced need to hire more agents—needed to be achieved in less than four weeks!

The Australian Football League solution

Because this was a short-term solution, and the lead-time was very short, Inference opted for a fully hosted service, with the AFL IVR directing callers to the speech system. Inference started work in the third week of August 2007, developing a prototype service based on the requirements provided by the customer. This was a fully operational natural language prototype.

At the end of that week, the human factors team worked with the AFL to test the solution with the various stakeholder groups, including customers, agents and management. As a result of this testing, the prototypes were refined and the grammars developed to a production quality. In the final week of August, the service was deployed onto the production platform and again tested with the various stakeholders and further iterated as a result of feedback from that testing.

This particular service was designed to deliver up to date information about football games, ticket availability and member entitlements that changed on a daily basis, so the AFL required some way to update the information delivered by the service real time. In this case, there was no back end database housing all the information required, so integration to an existing data source was not possible.

The solution delivered by Inference incorporated backend databases along with a simple web front end to enable the AFL to update the database as and when required. This was crucial, as the database was the source of much of the real time information that was read out in prompts at various dialogue states in the application. The rapidly changing information used TTS to read out those prompts, whereas the static prompts used recorded audio prompts recorded by 'the voice of the AFL.'

The resulting service therefore used recorded audio for all those prompts that were not going to change throughout the finals series and then TTS for those parts of the service where the prompts and information delivered to the caller was dynamic. The grammatical inference technology also allowed the AFL and Inference to tune the service very quickly and easily by identifying out-of-grammar utterances at regular periods and feeding them into the AI engine and redeploying an optimally spanning grammar set.

The Result

The natural language service providing all the information described above was designed and implemented within two weeks and delivered 88% full automation on Day 1. This automation rate was achieved even through callers had the option of being put through to an operator on request and were automatically transferred to an operator on two "no matches." From a business perspective, the AFL was able to handle the extreme demand on their phone service during this period seamlessly which was in stark contrast to previous years when callers had reportedly waited on the line as much as 60 minutes to get the information they required.

One of the key risks for the AFL in implementing this service was any potential damage to their brand as a result of poor customer feedback. The AFL callers are not generally early adopters of technology, and they are very vocal if they are not happy with the service being delivered. Remarkably, no customer complaint was received regarding this automated Finals Information Service.

For the AFL, an unexpected benefit of the service was the access to information about their callers and their reason for calling. The AFL was able to track which callers were calling, for which teams, and about which games. So, rather than the case in previous years when the AFL did not even know how many people were waiting in queues to have their calls answered, in 2007, not only were the calls answered immediately, they also learned who was calling and why.

How Inference Communications achieved this result

Inference leveraged the power of the grammatical inference engine to design, develop and deploy a suite of pre-packaged hosted services specifically for customers requiring short-term solutions with dynamic data. What does this mean? It means speech recognition services that are mixed-initiative, rely on changing back-end data, and are only required for short periods (along with the option for self management by the customer) can be delivered using Inference's technology. Services deployed using grammatical inferences automatically have the following attributes:

- *Mixed initiative:* Using grammatical inference the developer simply has to enter a small number of sample phrases that the caller might utter into the engine, and the development environment will automatically generate the grammar set with the coverage required for optimal accuracy.
- *Robustness:* The system includes automatic fall back to directed dialogue and DTMF.
- *Dynamic grammars:* The system generates grammars automatically when updated.
- *Mixed prompts:* The system can be deployed with recorded audio and TTS as required.
- *Web-based updating:* The availability of a simple Web interface for full content management and tuning.

The AFL Finals Service was possible because of Inference Communications' ability to generate sophisticated mixed initiative solutions with dynamic grammar generation in a very short time and at a price point that made the option commercially viable. This unique benefit conferred on Inference by the grammatical inference technology starts with the design process where the designers and human factors experts do not have to consider the implications of grammar development on their design, and they therefore have greater freedom to design for optimal service. The usability testing carried out by the design team is very fast and very 'real' because the testing group is interacting with a series of real prototypes of the types of systems they are seeking to deploy.

Summary

In short, the AI grammatical inference technology confers advantages to the speech application developer (and ultimately the customer) across all stages of the speech recognition application life cycle. From the ability to develop prototypes very quickly and allow the design team to operate more effectively and efficiently, to the actual development of production quality solutions with grammar files automatically generated to achieve optimal accuracy levels, and then throughout the tuning and maintenance phases. This is in addition to the ability to develop complex speech solutions in a time frame and at a price point that makes them a viable option for short-term solutions.

Next-Generation IVR avoids first-generation user interface mistakes

Bruce Balentine

Bruce Balentine, executive vice president and chief scientist, EIG Labs, Enterprise Integration Group, discusses the lessons on what to avoid taken from "first-generation" speech IVR solutions. Bruce specializes in speech, audio, and multimodal user interfaces. He has been designing user interfaces for more than thirty years, including twenty years in the speech industry. Balentine has several publications, including "How to Build a Speech Recognition Application" which is considered the definitive style guide for IVR human factors. The Third Edition of that book is soon to be published by ICMI. His latest book, "It's Better to Be a Good Machine ..." was released in March 2007. Balentine received his undergraduate degree in 1971 and his Master's in 1975 from the University of North Texas with work in electronic music and multimedia.

Introduction

I just returned from an extended tour of call centers in Australia and New Zealand, and I see some remarkable signs that the pendulum is swinging in that faraway region—just as we're seeing it swing in North America. The pendulum I'm talking about is the design philosophy that led to poor IVR design in a number of first- and second-generation systems. The kinds of user backlash that have led to phenomena such as the Paul English website, Saturday Night Live spoofs, and Wall Street Journal editorials are fading into the past. They are being replaced by a new generation of IVR systems that are faring much better.

In the Australia/New Zealand market, we see many earlier systems that suffered from three very specific bad practices:

1. Failure to integrate speech and touch-tone;
2. Over-emphasis on personae; and
3. Spending on statistical language model (SLM, "natural-language") open dialogues in cases where directed dialogue is preferred.

These practices have been on EIG's list of *Top Ten Reasons IVR Projects Fail* for the past several years, and with good reason. Together they have contributed to a general perception of telephone-based self-service as an enterprise-centric indulgence rather than a businesslike professional practice. Here I'd like to discuss each one briefly.

Speech and Touch-Tone

Users choose speech or touch-tone for all kinds of reasons, as shown in the following table. Indeed, we observe in usability tests that the same caller often switches between one modality and another *even within the same session*. The modality decision depends on the immediate context of the conversation. Neither modality is intrinsically "superior" to the other.

Why Users Choose Speech		Why Users Choose Touch-Tone	
Do things touch-tone can't do easily	Names, addresses, alphanumeric Complex transactions Multi-slot-filling	**Privacy**	Account number and PIN Don't want others to overhear conversation
Improve ease of use	Less taxing on human memory Keypad small, keypad in handset It's dark	**Environment**	Noise Distraction
Speed of entry	Talking is sometimes faster than touching Speaking from memory is convenient and easy	**Convenience**	Numeric data Desktop telephone with easy access to keypad
Accuracy	Speaking while reading rather than switching back and forth	**Accuracy**	Speech degrades in noise Error recovery dialogues frustrating or missing

In other words, the preference for one modality over another is not a characteristic of a given user's personality or preference—it is the result of the changing circumstances of the call. This is why early systems that simply asked the user at the start of the call, "Do you want to use touch-tone or speech?" were so ineffective. The user must have the opportunity to switch modality at any point in the dialogue. And it must be the caller's decision.

Over-Emphasis on Personae

Personae are "personalities" that designers give to an IVR. Some of the larger enterprises over the past decade became quite enamored of personae and devoted a great deal of time and resource to getting them right. Such enterprises valued the branding and the high-tech/high-touch characteristic of an especially natural and conversational user experience. Many succeeded.

But others focused on personae while neglecting usability. In these cases, the user experience was delighting as long as the dialogue never hit a snag, but the persona would often backfire when the TUI conversation went awry. In some cases the systems were simply incapable of recovering even common errors, leading to user interfaces that tried to be "fun" without being competent.

In some cases, careers were damaged as enterprise decision makers overspent on the personality and social properties of the user interface without considering user context. Many of these systems made negative headlines—not helpful for overall user acceptance of speech recognition.

Open Dialogue versus Directed Dialogue

Statistical Language Modeling (SLM) technology supports so-called "open dialogues." In this specialized use of speech recognition, users describe their problem in their own words.

In the following illustration, the horizontal (x) axis represents the most common reasons for calling, while the vertical (y) axis shows the percentage of all calls with that reason. The curve is typical—many calls are for the most common reason, with a rapidly diminishing slope as we move to the right. It is not unusual for the second or third bar—the second or third most common reason for calling—to be the "everything else" category.

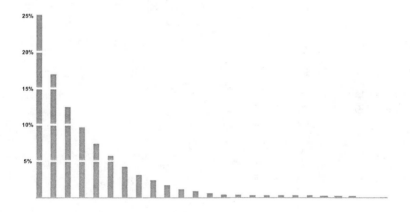

The problem with this distribution is that the SLM does a good job with the call reasons toward the left of the chart, but performs less well as we move to the right. The reason is that we have fewer and fewer samples of speech in our statistical corpus with which to train the SLM. What's more, the number of samples that we need—and hence the cost of the

SLM—grows exponentially if we want to model more than a handful of call reasons.

The upshot can be found in Pareto's law of maldistribution—we can handle 80% or so of the reasons for calling using a simple menu—at a fraction of the cost. To model the less-common reasons, we must spend and tune and spend and tune—often beyond the value of the end result. Australian markets in particular were bemused by "natural" systems that only responded if you "speak a certain way"—the layman's way of describing an SLM with limited topic coverage.

How did this happen? The core problem was that marketing homed in on "natural" rather than on "un-coached" as the ergonomic principle. SLMs are very handy as an alternative to directed dialogues—which take time to disclose to callers. But to have value, an SLM must support many call destinations *with approximately equal distribution*. Otherwise a standard menu will do the job.

> **The core problem was that marketing homed in on "natural" rather than on "un-coached" as the ergonomic principle.**

The Next Generation

The good news is that a new generation of effective IVR systems is beginning to appear. These IVR systems are based on past learning and early market mistakes, and so are much more effective at serving callers. They:

- Are designed for usability, not user delight;
- Exhibit much more robust error-recovery;
- Support both speech and touch-tone throughout, with intelligent cross-degradation between them;
- Are designed for change management by business owners; and,
- Offer dynamic broadcast messaging and contingency features for flexibility.

These and other features add up to create an integrated design that optimizes for all of the competing business goals: self-service, call handling time, skill-based routing, front-end data capture and transport for maximum agent efficiency, and total cost of ownership.

The enterprises in Australia/NZ are focused now on business goals rather than championing technology. As a result, we can expect to see significant uptake of speech recognition technology in both of those countries.

Multimodal Customer Service Transactions
Matt Yuschik

Matt Yuschik, Ph.D., Senior User Experience Specialist, Multichannel Self Care Solutions, Convergys Corporation, discusses how multimodal interactions can extend the voice interface into a different form of "natural language" interaction, where skilled agents can test the interaction before expecting the customer to do it. Dr. Yuschik designs and evaluates multimodal user interfaces for call center agents and for customer self-care devices. Matt previously designed and brought a voice-activated voice mail product to the market in Europe and the US. His designs are intuitive and easy-to-use, based on task analysis and turn-taking principles of human behavior. Matt has numerous patents and publications in the field of speech technology, voice activation, and multimodal interfaces.

Call Center SMEs

There is a compelling need to provide viable multimodal-self service transactions for customers who dial a contact center for service. New devices can support almost any user interface – text, graphics, voice, touch, etc. One way to migrate transactions from agent-assisted call center interactions is to develop easy-to-use end-user services. To start this process, we must first understand the caller behavior in customer service centers. Currently, contact center agents, subject matter experts in their own right, isolate a customer's problem to one of a set of specific solution by using a set of rich database search tools. Convergys considers our agents as problem-solving "solutioners," and draws upon their experience to help assess end-user needs, and to test multimodal services for our customers. This strives towards the goal that customers should be able to complete their own transactions on the device of their choice, and be highly satisfied with the result and experience.

User Interface Versions

To appreciate where multimodal transactions fit into the spectra of voice-enabled services, a narrow perspective on the history and evolution of human interactions at contact centers can be defined.

Version 0: All calls go to live agents. This is the original contact center process when a company handled every call personally. The agents used well-rehearsed scripts which they followed to consistently handle typical customer issues.

Version 1: DTMF (touch-tone), in the form "For X, Press 1", gave callers a way to take control to resolve simple, straight-forward issues. The menu structures began from agent scripts, and choices were grouped in a tree structure for navigation using the buttons of the telephone keypad. These menus forced the caller to take part in the "solutioning" process as well as follow the menus. Agents still handled difficult or uncommon issues.

Version 2: "Voicify" the DTMF prompts to the form "For X, Press or say 1." This is the first foray of speech into the dialog, but is essentially a voice overlay onto Version 1. It continued the strong coupling of the transaction with the DTMF menu structure with the belief that an option would match the caller's need.

Version 3: Initiate a Directed Dialog, like, "Please say listen, send or mailbox options." This avoids the mental mapping of numbers for choices (navigating menus), and leverages a VUI designer's skill to present logical choices that supports a flexible problem solving process. The most common use cases are handled through voice-enabled automation, with the agent available to resolve other issues.

Version 4: Encourage Conversational Language in response to an open-ended question, like, "What would you like to do?" The response drives the transaction by following the user's lead, versus imposing the menu structure imposed by automation. The interaction occasionally drops-back to *Version 3* as a way to provide options and guidance to the caller to move the dialog forward. There is a risk is that the caller's intention is not understood or the issue is not able to be resolved using the service. Once again, the agent comes to the rescue.

Version 5: A Multimodal approach that visually displays data and options, and expects a verbal response when it asks "What else?" Generally, voice is used for input, and text / graphics for output. The caller has complete control, though the system provides a number of ways for the caller to search for resolution of an issue, starting with vague terms and then refining the search to a specific issue.

Besides adding other modalities to the transaction, this sequence of *Versions* also shows an evolution from the agent's view of issues to the caller's view, from a highly structured decision-tree approach to an open-ended, flexible structure -- which can fall back to a more restricted structure that narrows the focus to a single issue. This approach provides a balance which lets the caller have flexibility to initiate the conversation in any way, and lets the automation suggest options when the caller hesitates, or data is needed to move the transaction forward. A back-and-forth dialog is maintained until all required info is obtained and issue closure can be achieved. This approach follows a flow driven by the caller yet guided by the automation. This interaction approach is more dynamic (more turn-taking) and less static (menu driven).

Call Center Transactions

Convergys has about 82 domestic and international contact centers that provide a rich opportunity to monitor agents and track caller interactions. These observations enable defining a model of agent behavior which contains: a solutioning part to isolate the key issue; an information gathering part with data entry and navigation through multiple agent screens; and, a resolution and closure part to present the solution. Even though some dialog is not directly relevant to the issue, it is permitted to increase caller comfort, acquire optional background information, or defuse caller frustration.

The Transaction Flow

Convergys constantly strives to improve self-service and increase customer satisfaction. It does this by observing incoming calls, where the various aspects of flow and pace in the dialog are clearly distinguished. By overlaying multimodal onto the existing GUI workstation in the agent environment, the agents have the flexibility to follow a problem-specific flow and use their voice to navigate between computer screens and populate a graphic interface. The agent converses with the caller to extract sufficient information and converses with the multimodal workstation to complete specific screen-based tasks. Maintaining a caller-focused transaction flow is a valuable method which leverages a multimodal UI.

The agent converses with the caller to extract sufficient information and converses with the multimodal workstation to complete specific screen-based tasks.

Certain tasks are easier to complete in one modality than another – speech is excellent for navigation, while graphics is better for presenting answers from database searches. Verbal shortcuts are created, and are spoken during the solutioning step to enable the agent to jump to screens where specific data is required. Caller data may be obtained at any time in the conversation and placed in a speech-enabled short-term memory until the information is required in a specific location of a GUI screen.

Transaction-specific flows are developed to emulate the steps taken by the agent and the caller, with the goal that eventually the callers will perform the transaction by themselves on hand-held mobile devices. The flows are first tested by the agents on their multimodal workstations. The flows also support backup and error handling should the solution veer off-track and require helpful redirection or restart. Only when the transaction passes the test of the agent using it for real-world transactions, is it considered robust enough for deployment on a smart handset. One limitation of multimodal services is the ability of the software on current mobile devices to support the services. This is a valid concern, but the introduction of more 3G phones and open API software enables multimodal applications to become more pervasive. Current technology encourages the use of mobile devices for more complex tasks, especially multimodal services.

Safety Net

Once an automated multimodal version of a contact center service is constructed, it is deployed in a limited pilot test with friendly users (generally, agents who use the device yet can fall-back to their existing workstation), tasked to validate success in a rich set of use cases. Usage patterns and results are monitored to identify and address any unanticipated pain-points. While the overall goal is to contain all calls, not all caller issues can be handled by automation, so effective intervention by an agent must be provided. A hidden agent procedure is followed when the caller starts an issue, and an implicit monitoring mechanism (viz., business decision rules) infers that the caller is having difficulty. An agent is bridged onto the call without the knowledge of the caller and receives the transaction history and context, as well as can listen to the caller's current and prior speech to move the transaction forward "behind the scenes." The agent intervenes only when needed, which may be for simple situations where the caller is difficult to understand, problem conditions that are beyond the capability of the automated solution, or when the caller's emotional state must be defused before a solution is attempted.

Future Work

Ongoing and future work at Convergys focuses on business sectors and use cases which are amenable to taking advantage of a rich multimodal environment. In the telecom sector, calls are received from customers experiencing service-impacting conditions. Whether it is the need to troubleshoot a set-top box, or download ringtones, the ability to show a video clip on the handset to step the caller to issue resolution has a huge advantage over talking the caller through a sequence of potentially complex steps. In the sales sector, a retailer can display a visual rendering of products (clothes, rental car models) using internet capabilities of a mobile device. Designing, prototyping and trialing these applications provide a rich opportunity to identify those tasks and transactions suitable for migration to the ever increasing number of 3G intelligent phones. The ability of Convergys to use contact center agents familiar with these transactions and willing to test alternative multimodal environments to solve caller problems is a very rewarding opportunity. Convergys is in a unique position to provide multimodal applications with cutting edge technology to match the behavioral habits of an increasingly technology-driven culture.

Listen and Learn: How Speech Applications Help You Understand Your Customers

Patrick Nguyen

Patrick Nguyen, CTO & Founder, Voxify, discusses how companies automated speech customer service systems can reveal insights on customer behavior not easily discovered by after-the-fact analysis of recordings. Nguyen is a veteran technologist with over 15 years of experience, Mr. Nguyen is an expert in speech applications, data warehousing, e-business analytics, and CRM. He has held technical management roles at Personify, McKinsey and Company, and Australia's Telstra Research Labs. Mr. Nguyen is a frequent speaker at the SpeechTEK, Voice Search, and Genesys G-Force conferences. He has authored articles on speech applications that have been published in contact center publications. Mr. Nguyen has a BS in Electrical Engineering from the University of Melbourne and an MBA from MIT's Sloan School.

Enterprises seeking to acquire and nurture profitable customer relationships rely heavily on data about prospects and customers to predict future buying intentions. While companies devote significant resources to the analysis of transactional data from web and point of sale systems, they have yet to utilize speech systems to obtain behavioral data that would be invaluable for modeling customer interests and tendencies. To date, speech analytics has been relegated to the contact center as a tool for improving agent or IVR performance rather than treated as a strategic technology for supporting business intelligence efforts.

Traditional online and offline sources of customer data have limitations on several fronts. Data collected from point of sale systems are *incomplete* because they record only the outcome of a purchase process. These transactional systems do not capture the shopping behavior leading to a purchase – or non-purchase – that is useful for segmentation and predictive modeling. Web systems capture some behavioral data, but can be *inaccurate* because the basic unit of measurement – a session – needs to be approximated from periods of user inactivity. Furthermore, the usage model on the web of multi-tasking and switching between multiple activities makes the measurement of engagement and response rates unreliable. For example, a web application that logs a five minute page view cannot determine whether the user has pored over the page contents or simply been distracted by another task such as reading an email.

A well-designed speech application, with its ability to capture every point in an interaction with a customer, can provide cleaner and more insightful behavioral data than the web and other customer channels. A few examples illustrate the unique analytic capabilities that speech applications can provide to marketing, product development, and other business decisions.

Measuring Promotion Effectiveness

Promotions, or "upsells," represent an important revenue generation opportunity for companies in

> "A well-designed speech application, with its ability to capture every point in an interaction with a customer, can provide cleaner and more insightful behavioral data than the web and other customer channels."

markets as diverse as retail and financial services. To quantify interest in a particular offer, a company needs to communicate the offer details to an audience and measure the response rate. On the web, the measurement of a user's reaction to specific promotional content is problematic due to uncertainty in the user's level of attention.

Consider a retailer of women's beauty products that promotes several subscription programs to deliver beauty supplies on a monthly basis. The retailer places promotional ads at various parts of its web site, but cannot determine if a user has noticed a specific ad to accurately measure its acceptance rate. Moreover, the retailer cannot assume that a user's lack of acceptance indicates consideration and rejection of the offer.

In sharp contrast, speech applications provide an excellent platform for playing promotions and gathering the analytical data required to study promotion effectiveness. Due to the special nature of a speech interface, the company can reliably determine a customer's reaction to individual offers. The speech application can measure precisely the number of callers who heard all or part of an offer. Furthermore, the speech application solicits explicit acceptance or rejection – and if the caller hangs up, rejection can be inferred. The data collected in this manner provides great predictive power about future acceptance rates for the retailer's various promotional offers.

Capturing Customer Intent

On the web, the common practice for avoiding invalid entries is to constrain choices by using controls such as check boxes, radio buttons, and combo boxes. This practice places the burden on the customer to convert a desired input into a valid choice, and in the process misses the opportunity to capture the original customer intent.

As an example, when a web user makes a payment by credit card, the supported credit card types are usually presented as radio buttons. If the customer's preferred card is not listed, he has to adjust his initial inclination by choosing another card. As another example, a user on an airline web site may be forced to pick a departure city from a list box that only contains the cities served by that airline. If a customer wishes to depart from an airport that is not listed, he has to adjust by selecting a nearby airport that is listed. In both these cases, the original desire of the customer is not recorded by the web application.

In a speech application, a customer's response cannot be restricted, and may deviate from the predetermined options that the application's designers anticipated. Long regarded as a VUI design challenge, this phenomenon nonetheless presents an opportunity to discern the customer's objectives in a way that is not possible through the constrained input mechanisms on the web. In the above examples, the customer's preferred credit card or departure airport would be captured (even if the customer is later asked to make a new selection), providing data that could be extremely useful for segmentation or other marketing purposes.

Understanding Buying Behavior

For many shopping processes, the web provides an efficient user experience due to its high information density and widely adopted norms for navigating a graphical user interface. From a data completeness perspective, however, this efficiency obscures the thought process that a customer follows to reach a purchase decision.

Consider the case of a customer shopping for a hotel room. In the typical web booking experience, the user enters the destination city and trip dates, and is then presented with a windowed list of hotel results. The user is able to browse through the results list to gain a general sense of locations, hotel chains, star ratings, room rates, promotions and other relevant hotel attributes. Although the user has the option to filter the result set, he is not forced to declare his interest along each hotel attribute. The rich graphical interface allows the user to quickly scan and mentally filter product choices in a way that the web application cannot record.

The same shopping function in a speech application can be analyzed in much greater detail. The limitations of the speech interface prevent the application from presenting long lists of detailed product information to the end-user. Instead, a VUI best practice is to reduce the result list to a manageable length by asking the customer to specify his search criteria (e.g., "Would you like to search by location, hotel chain or price?"). The customer's answers to these filtering questions reveal which attributes are most important to his purchase decision. This behavioral data is a valuable by-product of the voice interface's need to break down a complex interaction into the discrete steps that expose a customer's thought processes.

The Required Speech Infrastructure

The advantages of completeness and accuracy enjoyed in the automated speech channel can only be leveraged if data about each call is captured and correctly interpreted. The most important data – and the data unique to speech – relates the customer's conversational behavior. Conversational behavioral data is both more abundant and more in need of interpretation than traditional customer profile data. It entails not only all the specifics of what the system said and what the customer uttered at each point in the conversation, but also what the customer meant by a series of exchanges. While even a hand-written VoiceXML application can be augmented to log low level events from the speech recognition, like caller utterance and recognition confidence, properly interpreting a series of events as a conversational behavior is much more challenging. For example, a customer uttering, "...no, I meant last Monday," must be correctly interpreted as correcting his response to the previous question rather than addressing the current question. The framework for making these interpretations is a vital component of an application designed for measurability.

Moreover, the system needs to be resilient in the face of the constant changes in phrasing, call-flow, and grammars that are required to keep an application relevant to the current business need. As the application evolves, the capture and interpretation of behavioral data must remain complete and accurate. This requires an analytics framework that can adapt automatically to changes in the VUI design.

Finally, the data must be structured in such a way as to inform the business intelligence analysis process. This entails casting the data in multi-dimensional databases and forms amenable to ad-hoc data mining analyses.

In summary

When supported by an appropriate infrastructure, speech applications can capture behavioral data that is more complete and accurate than the customer data traditionally available from web and point of sale systems. Armed with this data, a company can build predictive models across its business lines to more effectively target prospects, convert prospects to customers, and increase customer lifetime value. As companies become adept at leveraging customer analytics for business insight, the strategic role of speech applications in providing this behavioral data could become as important as the ROI generated by automating service functions.

Too Much of a Good Thing

James A. Larson

James Larson, Speech Applications Consultant, Larson Technical Services, discusses some problems with "protecting" customers from live agents. Dr. Larson is an independent consultant and VoiceXML trainer, co-chair of the World Wide Web Consortium's Voice Browser Working Group, and author of The VXML Guide. Jim can be reached at jim@larson-tech.com.

Many customers prefer automated agents to live agents because live agents sometimes speak with difficult-to-understand accents and occasionally give incorrect information. Mostly, customers prefer automated agents because the customers are not placed on hold and forced to listen to boring music, or worse, radio advertisements. Automated agents answer the phone quickly, are relatively easy to understand, and provide consistent, accurate information. These are all good things. Automated agents can provide good self-service.

But too much of a good thing is not a good thing! Too much chocolate cake results in an upset stomach. Forcing customers to always deal with automated agents results in upset customers. Customers are not dummies—they know when automated agents can help and when live agents can help. Forcing customers to speak with automated agents without allowing customers to speak with live agents is aggravating for the customers and a waste of connect time for both customers and automated agents.

> **Customers are not dummies—they know when automated agents can help and when live agents can help.**

Remember how customers loved to hate DTMF calls? Many customers believed that DTMF agents stood in the way of connecting with live agents. DTMF agents were often perceived as gatekeepers, roadblocks, and symbols that the company didn't care about its customers. The phrase, "your phone call is important to us," has come to mean exactly the opposite.

Voice user-interface designers have long understood that forcing users to interact with automated agents is counter-productive. *"Always enable customers to escape from automated agents"* was one of the ten top guidelines identified by VUI designers during a day-long think tank.[2]

---

[2] http://www.speechtechmag.com/issues/9_7/cover/11311-1.html

Vocal Labs' Peter Leppik[3], conducted 60 usability experiments. The resulting data strongly confirms two important principles supported by most voice user-interface designers:

Principle 1: Forcing customers to stay with an automated agent doesn't really help customers achieve their goals.

Figure 1 shows the results comparing automation (where customers are able to complete their tasks) with frustration (how difficult is it for customers to reach a live agent). Leppik found that there is a correlation between automation and frustration, but not a strong one. Making it difficult to reach a live agent might increase the number of successful self-service transactions, but the effect will probably be small.

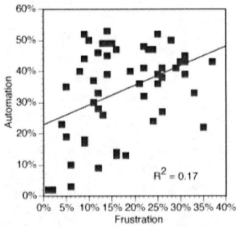

Figure 1: As the frustration rate (difficulty of reaching a live agent) increases, the automation rate (number of self-service tasks successfully completed) may increase marginally.

Principle 2: Forcing customers to stay with an automated agent increases customer frustration and decreases the successful completion rate.

Figure 2 illustrates the results of comparing the frustration rate with the completion rate. Note that an increased frustration rate correlates strongly to a drop in completion rate. Figure 3 shows an increased frustration rate correlates to a dramatic drop in satisfaction.

---

[3] http://www.vocalabs.com/resources/newsletter/newsletter22.html#article1

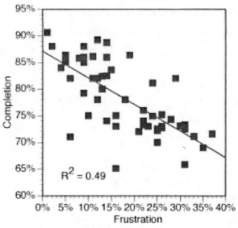

Figure 2: As the frustration rate (difficulty of reaching a live agent) decreases, the completion rate (number of self-service tasks successfully completed) increases.

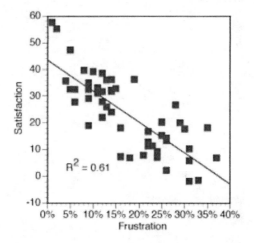

Figure 3: As the frustration rate (difficulty of reaching a live agent) decreases, the customer satisfaction level (number of self-service tasks successfully completed) increases.

Customers often choose which company they do business with based on the company's service. Companies can choose to support any of several customer-service strategies:

- *Customer-friendly*—Upon customer request, connect the customer to a live agent. The company will invest dollars to hire and train live agents. The company's investment will likely recoup the company investment as repeat customers receive the service they want.

- *Customer-hostile*—Never let customers connect to a live agent. The "discount" company will continue to serve thrifty customers, but many top-paying customers will go elsewhere because of poor customer service.
- *Customer-ranked*—the most valuable customers receive first chance to connect to a live agent; while less valuable customers wait. This is like first-class vs. second-class seats on airplanes—you get what you pay for.

How will your customers judge your company based on customer service? Are you using the right customer-service strategy? Are you forcing your customers to use too much of a good thing?

You Can't Think of Everything: The Importance of Tuning Speech Applications

Ian Guinn

Ian Guinn, Speech Application Developer, LumenVox, discusses the necessity and power of tuning an application and provides examples. In addition to assisting with the development of LumenVox's core speech recognition technologies, Ian has developed several speech applications for the company, including many of its flagship demonstration speech applications. Additionally, he consults with clients to troubleshoot recognition and design issues.

An old proverb says, "Man plans, God laughs." Nowhere is this more true than in software development. As developers, we try to plan for every contingency. In our minds we walk down every path our users may travel and compensate accordingly. Without fail, customers will traverse paths never intended for a user's feet. Immediately they come back to us and say, "Look, it's broken!" No matter how much we plan, we must be able to modify our systems based on how our users actually use them, not on how we expect them to be used. In the world of Voice User Interfaces, this process is called tuning, and a tuning tool is a vital component of VUI development.

Without fail, customers will traverse paths never intended for a user's feet.

A good VUI is difficult to write and test. When users complain that an application is "broken," it may mean that the speech recognition is failing, but just as often it means that the developer of the application simply did not take into account the sort of responses users would give. Unlike DTMF applications—or even traditional GUI applications—an application with a speech-driven VUI allows for an endless number of potential user responses to any given prompt.

As an example, LumenVox produced a VUI demonstration that would illustrate the capabilities of our Speech Engine. The demo would tell people the current weather for a city of their choice. Using U.S. Census data, we built a grammar with the name of every city with a population of more than 5,000 people. Once a user selected a city and state, the system retrieved weather information from the Internet and read it to them using text-to-speech.

It was a fairly straightforward design, with no obvious snags or hang-ups. Almost immediately after the system was deployed, however, users reported it was failing. We immediately called the system to ensure it was working:

> **Speech Application**: *Please tell me the area you would like the weather for.*
> **Caller**: *San Diego, California.*
> **Speech Application**: *I heard "San Diego, California." Is this correct?*
> **Caller**: *Yes.*
> **Speech Application**: *The weather for San Diego, California is …*

The voice interface we had designed seemed to work fine from a technical perspective. The speech recognition was accurate, and all the components were working together as expected. And yet users kept reporting the system was failing. It wasn't until we reviewed actual recordings of calls using our Speech Tuner that the problems in the system's design were exposed. Our Speech Tuner allows us to listen to the audio recordings of callers, see what the Speech Engine recognized, and see how changes to grammars would have affected recognition.

One key feature of the Tuner is its *Call Browser*, a module that allows us to see details about a call and each utterance in that call. This way we can follow a user through a call, see what the caller said, and see what the Engine recognized the response as. A common user experience went like this:

> **Speech Application**: *Please tell me the area you would like the weather for.*
> **Caller**: *92123.*
> **Speech Application**: *I am sorry, that is not a valid choice. Please try again.*
> **Caller**: *ZIP code 92123.*
> **Speech Application**: *I am sorry, that is not a valid choice. Please try again.*
> **Caller**: *Area code 619.*
> **Speech Application**: *I am sorry, that is not a valid choice. Please try again.*
> **Caller**: *The moon, or anywhere nearby.*

We listened in horror as the users ripped our robust application to shreds. While a developer cannot plan for every possible phrase a user may utter, it was clear the prompt was misleading our callers. The seemingly simple request of "Please tell me the area you would like the weather for," was far too open-ended. We heard responses such as "Near my house" and "The beach."

Needless to say, after reviewing the results of our seemingly simple and fail-proof application, we decided that the initial prompt needed to be changed to something that elicited a specific response instead of such an open-ended question. We changed the prompt to say, "Please tell me the city and state you are interested in," and the application's success rate improved significantly. By reviewing actual calls with the Tuner, we were able to quickly pinpoint the exact cause for user failure, and adjust the system accordingly. Just as importantly, we were able to review the results of the change to ensure it was successful.

The next demonstration application we developed was a fake pizza-ordering application. This demo allows users to choose the toppings, size, and crust of a pizza. As with the weather demo, we built the initial application, tested it internally, and deployed it. Once again, users immediately complained that the application simply did not work. When the system asked users what size pizza they wanted, we expected them to ask for a small, medium, or large pizza. Listening to calls, we heard interactions such as:

Speech Application: *"What size pizza would you like?"*
Caller: *"Twenty-seven inch, please."*
Speech Application: *"Hey, we only make three sizes of pizza: small, medium, or large."*
Caller: *"Medium, then."*
Speech Application: *"Was that a medium?"*
Caller: *"Yes."*

Even though the mistake was handled by the "no match" prompt, this failure to answer an ambiguous question the way we expected could lead to caller fatigue and frustration. This condition can occur if callers encounter just a few questions that don't move them forward to the implied goal of the system (in this case, ordering a pizza).

Using our Speech Tuner, we were able to provide immediate and satisfying proof of the need for a change in the system. We added a grammar that would accommodate users specifying a pizza size in inches, as well as being able to say "small," "medium," or "large."

Unlike the weather system, we decided to add the grammar to accommodate a larger range of responses instead of simply rephrasing the prompt. In the case of the weather demo, there was no reasonable way to accommodate requests like "I need weather near my house." But the responses to the pizza demo were within a limited domain that could be easily handled by a modified grammar, so it made sense to make that change.

Before we made the changes to our live application, we needed to test the grammar. To do this, we transcribed a large number of utterances—the Tuner's built-in transcriber aids us in this by automatically entering the Speech Engine's result into the transcript, but for out-of-grammar utterances we still need to spend time transcribing what the users said.

Once we had transcribed call data, we were able to make use of the Speech Tuner's *Grammar Tester* component. The tester takes transcribed interactions and gives us a list of the grammars that were active during the recognitions. It also gives us a wealth of statistics about recognition accuracy, based on the transcripts and the recognition results.

The key feature of the tester is the ability to modify the grammars and then run the audio through the Speech Engine, getting new results based on our new grammars. This allowed us to evaluate how the application would handle our original responses with the new grammar entries (the ones that allowed users to specify a size in inches). We saw our semantic error rate drop significantly, because our grammars now accommodated what the users were saying. The grammar test provided us with empirical data to show that the change would be a beneficial one that could be done swiftly and without negative user impact.

Using a good tuning tool allows developers to quickly harness the *only* experiences that really matter: those of users. Only when we understand how our users actually interact with our speech applications can we then plan improvements. And, with effective testing tools, we can accurately assess how those changes will affect our applications before deploying them in production environments.

Speech Recognition, the Brand and the Voice: How to Choose a Voice for Your Application

Marcus Graham

Marcus Graham, CEO, GM Voices, discusses the implications and some guidelines for choosing a voice and a delivery style for telephone dialogs. In 1983, Marcus started producing professional voice messages for telephone systems. Using theatrically trained voice talent and actors, music and special effects, the company today produces studio-quality audio programs for telephone systems, automotive navigation systems, multimedia CDs, and Internet web pages Instant communications technology also allows them to translate and record in 60+ languages from cities around the world including London, Toronto, Tokyo, and Frankfurt.

What began some thirty-odd years ago as an effort to let callers know that your business was closed, has grown into something far more sophisticated and critical to businesses today. Those simple answering machine messages such as, "Our office is closed..." have morphed into a wide range of automated call routing, information dispersing, and transaction completing technologies that are changing the way our society does business.

As more companies implement voice-driven self-service applications to lower cost, the smart ones have already recognized the impact this vital contact channel has with their customers. A critical part of the customer experience in automated voice applications is the recorded voice that guides the caller to their desired information or transaction.

Historically, the voices heard on a company's recorded telephone messages were given little thought because they were originally considered an extension of the receptionist. The receptionist answered the phone live during the day, and after closing time the receptionist's recorded voice answered via a phone answering machine. When touch-tone automated attendant and IVR systems arrived in the 80s and 90s, the receptionist continued doing many of the greetings and menus.

While the receptionist continued providing the voice on some applications, radio announcers began recording more of the messages. These 'enunciation' experts could pronounce any particular word perfectly, but they didn't sound natural. They had that 'disc jockey' sound. In fact, most automated systems working today use this sort of voice talent.

Let's assume that the speech application is functionally sound with effective call flows, scripts, and high recognition rates. The only real aspect the caller 'connects with' then is the voice. They know it's not a real person, but they do want it to be personable. Research has shown that people attribute human characteristics to speech applications because it just seems natural to do it. After all it is a voice speaking to them, right? It stands to reason that the more pleasant the voice, the better the exchange.

Speech Recognition

When speech recognition began to make its way out of the labs and into the marketplace in the late 90s, it soon became apparent that encouraging callers to speak naturally for better recognition rates was largely accomplished by speaking naturally to them with the prerecorded voice prompts. This realization dramatically raised the level of quality expected for recording greetings, menus, and prompts for the telephone.

> **Encouraging callers to speak naturally for better recognition rates was largely accomplished by speaking naturally to them with the prerecorded voice prompts.**

Ironically, at the same time, much of the higher end advertising and production world began moving toward using more believable voice actors in television, corporate productions, and shows. Using actors posing as 'real people' became a more credible and effective communications technique that has become even more widespread today. Most of the successful speech applications in use today employ actors as the voice talent.

An important part of creating a virtual personality for a speech application is developing a persona or biography to clarify the impression the company wants to leave with the caller and for use by the actor recording the messages. Who is this person answering the phone? How old are they? Where do they live? Do they have kids? Using this biography, the voice actor can become the person who's answering the phone and maintain consistency throughout the recording of hundreds or even thousands of voice prompts.

The Brand

How do you go about identifying the type of voice that will work best with your application? It starts with the brand. Think of the brand as a container that holds every experience the user has had with that company or product. All the good, bad, and the ugly touchpoints paint a picture in the individual customer's mind. The voice that is ultimately chosen for the company's speech application should embody the same attributes as the brand.

With multiple applications, there may be a need to brand by department, depending on the company and requirements. For example, specialized speech applications for particular departments may warrant positioning for something other than the overall brand and may require a different personality or strategy.

The automated-voice application is simply another contact channel with customers. It should be given the same care and consideration as other contact points. Voice customer self-service has become one of the dominant contact channels for many companies. As a result, Chief Marketing Officers have discovered that it has a huge impact on the customer's brand awareness.

The good news for people on the technical side is that the marketing department needs to pony up some budget money for implementing speech recognition applications! We've heard numerous respondents in focus groups and in-depth individual interviews state that the choice of speech recognition vs. touchtone is a competitive advantage. Positioning speech as a marketing tool may ease the financial burden faced by many IT departments if they play it right within their organization.

The Voice Solution Provider

The discussion of the voice talent selection usually goes from considering 'the brand' directly into 'choosing a voice,' but a more fundamental decision should be considered selecting the voice talent. That decision has a large impact on the final product: You need to evaluate the ideal way for your organization to *get* those voice prompts. There are three basic sources: 1) you, 2) a technology provider, or 3) a voice solution provider.

You

You manage the voice talent recruitment, audition, selection, contracts, and general management process. This is rarely an effective solution. Just as your business specializes in its area of expertise, you've got to decide if you want to learn this business yourself or hire people who do it everyday. On occasion, we've had companies that had a relationship with a voice talent and requested that we work with them on their behalf.

Technology Provider

This is a one-stop-shopping solution that many companies choose. A technology provider, such as a Nuance, Intervoice, Genesys, or IBM, has a professional services group with experience in this area and can do an effective job. These companies often outsource to a company in the next group to provide this service.

Voice Solution Provider

The third option uses a specialized provider, such as my company, GM Voices, that focuses entirely on providing voice talent solutions for speech applications. These specialized companies typically have more voice talent options, quicker turnaround, lower costs, and more natural-sounding concatenated speech (account and phone number strings). This option can help you with broader voice-branding solutions by putting your selected voice talent on all the other automated applications such as automated attendant, call center queues, and even on-hold messages.

The Voice Talent

Once you've decided the best option for your organization, then you need to look at voice talent. The choice you make in the section above will impact this process because each provider will likely have options readily available to you relating to voice talent. Regardless of which direction you go, you may consider an in-house voice or an out-sourced voice.

In-House Voice

This is usually a throwback to the after-hours message recorded by the receptionist many years ago. I know of a few staff members who turned auto-attendant work for their company into a career as a voice talent. But realistically, you can't expect a quality application from non-professional voice talent. Familiarity with voice-over recording, maintaining performance consistency, and a dozen other considerations are not natural. These skills are learned during years of refining the craft in voice acting.

Outsourced Voice

These are professionals who can deliver on most of the requirements. Of course, there are many professionals who 'won't pass the audition' for various reasons. Remember, choosing a voice is a very subjective process. It's hard to articulate why you like a voice or don't care for one, but you know it when you hear it. The difference between a Radio Announcer, Voice Talent, and Voice Actor is worthy of some discussion.

A Radio Announcer is typically a rich-voiced performer who's been in the radio business for some time. Male and female announcers are heard everyday all over the country. Next time you listen to the radio, listen to the words. They're usually pronounced perfectly and they likely have a deep tone, but they don't sound like real people. It's very difficult to turn off the patterned cadence of radio talking. "That's right, now back to more music." It's hard to say that like a real person.

A Voice Talent is someone who's trained in using their voice. They sound somewhat conversational, but it's still not quite real. These people usually are the anonymous voices you hear on radio and television commercials. They also do much of the narration you hear in business presentations or even on television. Anyone who's ever done a voice-over will call themselves a voice talent. It's really hard to articulate the difference.

A Voice Actor is an *actor*. They've likely been trained in drama or theatre where they learned how to 'become' the character in the show. It's about sounding real by being real. That's why the persona biographies are so important. That's how they figure out how to perform when the microphone is turned on. They are the guys and girls next door who talk to you like…the guy and girl next door!

When it's all said and done, you can't put a voice talent on a spreadsheet. Choosing a voice is still a very subjective process. And there are clearly preferences in a particular application for gender, age, and other factors.

One factor that will play a growing role in choosing voices is research. We've worked on projects for large IVR users recently where the marketing players are participating in the discussion and enthusiastically embracing traditional consumer-based research to validate voice choices. Through focus groups and in-depth individual interviews, we're finding out from customers what they really want. Marketing leaders will continue to get more involved in large-scale speech implementations as they increasingly impact the brand in the mind of the customer.

Beyond Best Practices, A Data-Driven Approach to Maximizing Self-Service

Joe Alwan and Bernhard Suhm

Joe Alwan, VP/GM, and Bernhard Suhm, Director of Professional Services, of the AVOKE Caller Experience Analytics business at BBN Technologies discuss a quantitative methodology to inform VUI design choices and maximize self-service. Alwan is responsible for the AVOKE Caller Experience Analytics division at BBN. Suhm manages the professional services team and holds several patents covering the methodology. Prior to joining BBN, Alwan was VP/GM of the contact center division at Empirix and has been working in enterprise technology solutions for 24 years. Suhm was a Senior Consultant with Enterprise Integration Group, a Senior Scientist at BBN, and has 14 years of experience in research and development of speech recognition systems. Alwan holds a B.S. with honors in electrical engineering from the University of Illinois. Suhm has a PhD in Computer Science from Karlsruhe University.

Perhaps nowhere are the limitations of best practices more evident than in the design of IVR user interfaces. Certainly every application is deployed with good intentions, using a consensus of best practices from its design team. Yet caller dissatisfaction is so widespread that there's an organized consumer revolt—with tens of thousands of visitors monthly to the "Get Human" website (www.gethuman.com), all 500 IVR cheats in a vcard file, and a service (www.nophonetrees.com) that will navigate the cheat for you and then ring your phone.

While best practice designs can certainly reduce dismal caller experiences and deliver minimally acceptable service levels, our research has shown that the unique characteristics of each contact center must be examined closely to optimize self-service and achieve excellence in customer experience. In this column, we'll examine five contact centers to illustrate a data-driven approach to voice automation strategy and design. The maximum possible self-service rates in these centers range from 41% to 65% - and they are all from the same industry! Our experience proves that benchmarks and standards alone offer little value, since center-specific factors typically outweigh commonalities within an industry (or even within a company).

Our findings are based on a patented methodology for measuring with statistical certainty the frequency distribution of the true reason for the call and the caller's end-to-end experience. The methodology applies to all inbound calls, whether handled by self-service or live agents. It is based on monitoring and dissecting the contact process from the moment the caller dials until they hang-up [Bernhard Suhm & Pat Peterson, "A Data-Driven Methodology for Evaluating and Optimizing Call Center IVRs," *International Journal of Speech Technology*, Volume 5, Number 1 / January 2002, updated version in *Human Factors and Voice Interactive Systems*, Gardner-Bonneau, Daryle; Blanchard, Harry E. (Eds.), 2nd ed., 2008]. We have had the opportunity to apply this methodology in contact centers across many industries, including financial services, transportation, healthcare, technology, communications, and energy.

Why Customers Call

Understanding why customers call is fundamental to any analysis of customer service, whether the goal is maximizing self-service or improving satisfaction. Customer needs must drive the design of automated interaction and agent handling, and these needs differ for each call reason and from one business to another.

Understanding why customers call is fundamental to any analysis of customer service

We obtain the distribution of true reasons for calls by analyzing the caller's IVR selections and the content of the agent dialog. To quantify the potential for improving self-service, we distinguish reasons that could be self-served—either by existing functions or with enhanced functionality—from reasons that must be handled by agents. Only those calls that are routine in nature and are handled by agents in a straightforward manner are candidates for self-service. Table 1 shows the distribution of self-servable calls from 5 utility customer service centers.

	A Region 1	A Region 2	A Region 3	B	C
Balance/Account Information	8%	12%	24%	14%	4%
Make a Payment	10%	12%	20%	6%	16%
Payment Extension	2%	3%	3%	10%	3%
Payment Arrangement	5%	4%	3%	10%	0%
Payment Notification	1%	5%	1%	4%	4%
Enroll in Payment Plan	1%	1%	0%	0%	0%
Meter Read	6%	1%	0%		
Start/Stop/Transfer Service	9%	13%	8%	6%	11%
Order Status	1%	2%	3%	1%	1%
Service Plan Info and Pricing	1%	0%	0%	0%	0%
Outage Report/Update	3%	1%	2%	4%	3%
Total Self-Servable Calls	*45%*	*52%*	*65%*	*55%*	*41%*

Table 1: **Distribution of Self-Servable Call Reasons on the Main Contact Number for Five Utility Customer Service Centers (percent of total volume)**

We can learn a lot about why customers call their utility just by studying the distribution of true reasons for the call.

- For all the utilities BBN has worked with, the average upper limit on self-service is 50% of all inbound calls. This upper limit includes self-service potential from new applications. Quantifying upper limits is crucial to determine how much room exists for improvement and to frame the business case for new applications.

- The most frequently requested self-servable transactions are: balance and account information, making payments, and other payment-related inquiries. Calls to start, stop or transfer service are next, though only portions of these complex transactions can be automated. Outages are one of the less frequent reasons on average, but self-service is crucial to handle the resulting volume spikes.

- Adding in the distribution of calls reasons for those that are too complex for automation, the total inbound volume can be classified into the following four major categories: billing and payment related inquiries; service orders and their follow-up; outages; and all "other" requests. Table 2 shows the average frequency for these top-level categories (19% of calls terminated early with no evidence of the reason for the call). Such a quantified distribution is the single most important consideration in the design of main menus and call steering strategies.

Billing and Payments	53%
Start, Stop or Move Service	16%
Outages	5%
Other	7%
Call purpose unknown	19%
	100%

Table 2: Top Four Call Reason Categories

These similarities across multiple utilities make it useful to reference an "average" utility, and that's what "best practices" can design for. But we also notice striking differences across utilities:

- *Upper limits of self-service:* The total self-servable volume varies dramatically from 41% to 65% of all calls. "A Region 3" stands out at the high end. This was attributed to: comprehensiveness of the existing self-service functionality; encouragement of customers to use it; ease-of-use; and less seasonal variability in weather. Weather variability translates into fluctuations in utility bills and more complaints—most of which must be handled by agents.

- *Alternative channels:* A separate payment contact number in "A Region 1" and "A Region 2" results in fewer payment-related calls to the main contact number. Further analysis can measure the effectiveness of multiple contact numbers and the corresponding implications for menu design.

- *Services offered:* "A Region 1" offers gas service in addition to electric service. This drives additional volume for: Meter Read, Outage/Gas/Light Questions; and Service questions—the latter due to appointments for starting or stopping gas service.

The difference in the distribution of call reasons in these five utility contact centers is just the beginning. More unique attributes become evident as we drill further into self-service utilization—all of which must be considered in the design of automated systems. Our experience shows that such variations exist across companies within every industry—and even across call centers within the same company.

Drivers of Self-Service Utilization

Many contact center initiatives are aimed at improving operational efficiency, which often means increasing self-service. Studying the success of automated dialog systems based on end-to-end calls provides a clear understanding of the factors that drive self-service success.

	A Region 1	A Region 2	A Region 3	B	C
Maximum Possible Self-Service	28%	28%	53%	34%	26%
Balance Information	✓	✓	✓	✓	✓
Make a Payment	✓	✓	✓	✓	✓
Payment Extension	✓	✓	✓	✓	✓
Meter Read	✓		✓		
Order Status			✓		
Outage Report/Update	✓	✓	✓	✓	✓
Full Self-Service Achieved	**10%**	**16%**	**43%**	**19%**	**8%**

Table 3: Self-Service Upper Limit With Available Functions, and Achieved Full Self-Serve.

It is widely assumed that functionality equals self-service. Additional functions certainly increase the maximum possible self-service. Table 3 shows which functions are implemented at each utility, and the resulting upper bounds for self-service. The upper limits were calculated by adding up the corresponding self-servable reasons from Table 1. The last row in Table 3 shows actual self-service achieved, which is clearly correlated with maximum potential. Indeed, functionality does matter.

...more self-service opportunities are lost to flaws in the menu structure and in the caller identification flow than to lack of self-service functions or their usability.

Usability matters also. Only well-designed self-service functions will allow callers to actually complete their intended task. But while usability is necessary, it is not sufficient for self-service success. In our experience, more self-service opportunities are lost to flaws in the menu structure and in the caller identification flow than to lack of self-service functions or their usability.

Navigation & Identification Success

Table 4 shows that the self-service conversion rate (the percentage of self-service achieved relative to the maximum potential) is strongly correlated with how many callers navigate the menus correctly, and how many callers successfully identify themselves in the IVR. The best system "A Region 3" converts 81% of its self-service potential into actual full self-service! This system is characterized by a menu structure that has 82% of all callers picking the right options, and successfully identifies 64% of all callers.

	A Region 1	A Region 2	A Region 3	B	C
Self-service Conversion Rate	34%	59%	81%	56%	31%
Navigated Menus Correctly	76%	67%	82%	59%	51%
Caller Identified Successfull	23%	30%	64%	51%	18%
Modalities	Touchtone	Touchtone	Touchtone	Speech	Touchtone

Table 4: Impact of Navigation and Identification Success on Self-Service

In our experience, high self-service utilization is not necessarily correlated with speech recognition. In these five centers, the speech-enabled system (utility "B") occupies a middle rank in terms of self-service conversion, while a touch-tone only application ("A Region 3") achieved the highest self-service conversion and utilization.

A more in-depth analysis of paths taken by callers reveals the factors that drive identification success: prompting as many callers to identify as is reasonable (given their stated reason for the call); leveraging ANI where applicable; and offering more than one way for the caller to identify.

Data-Driven Roadmap

The final step of our methodology translates new insights into actionable priorities. As the data has shown, a unique set of priorities emerges to maximize self-service for each of the five centers we've examined. Utility "A Region 3," the best performing center, could focus on automating portions of start/stop/move service calls. A Regions 1 and 2 should concentrate on identifying more callers successfully. And centers B and C need to focus first on menu navigation improvement.

When fully applied, our methodology also quantifies each improvement opportunity in terms of expected reductions in agent talk time and reductions in the number of callers with negative experiences. Based on extensive analyses of their actual callers, our customers have confidence that these benefits are achievable and are able to build enterprise support for funding new self-service initiatives.

Getting The Data

Studying caller needs and self-service potential systematically in this fashion—which we call the AVOKE Caller Experience Methodology—

requires visibility of the caller's entire experience from dialing to hang-up. In most centers, this data is scattered across systems, centers and business partners—making such analyses prohibitively complex. To overcome this obstacle, BBN has developed the AVOKE Call Browser system. Provided as a hosted network service, the AVOKE Call Browser system records complete end-to-end calls and extracts metadata describing the caller's experience using proprietary audio and speech recognition technologies. The results are provided in near real-time in a secure web application. In a matter of weeks, contact centers get unprecedented end-to-end visibility of customer experience without having to install any new hardware or software.

Simply "say what you want"—not so simple

William Meisel

"Just say what you want and get it" is a simple user manual for a speech interface, but difficult to live up to. "Keep It Simple, Stupid," the "KISS" principle, is a common reminder that we often can outsmart ourselves by being too clever, or more accurately, creating an overly complex solution when a simpler one will do.

The point is well taken in speech solutions. KISS would seem to apply, perhaps modified to Keep It Short and Sweet. Just say what you want and get it. Perhaps KISS should be considered a basic principle of speech interactions.

A customer calling a contact center just wants to solve a problem quickly, and asking for a lot of information that the customer knows won't be relevant to their inquiry can be both annoying to the customer and lengthen the call unnecessarily. Speech technology should not be a repetition of "touch-tone hell," where the customer is forced to navigate a series of often-confusing menu options.

A mobile phone owner may just want to find a feature on their phone quickly and not a display of a series of menus or icons—the visual equivalent of touch-tone hell. Users of mobile data services just want to get to a site or service quickly, and typing a web address or search term can be clumsy on a small device. Further, on a small device, a search result in the form of a long list is less useful; a short dialog—the norm in speech interactions—can disambiguate the request and give more targeted options.

But maintaining simplicity on the surface is not a simple task. "Natural-language" call routing required the development of sophisticated statistical methods both for speech recognition and for interpretation of the results of the recognition in order to take the correct action. Just saying an address to get driving directions required not only conquering a difficult speech recognition task, but also assembling a database that was in a format that speech recognition technology could use ("Dr" interpreted as "Drive," for example.) Achieving simplicity on the surface often requires a complex effort in both design and processing that the user observes only in the results. Web search using text input is one of the most successful examples of surface simplicity backed by hidden complexity, perhaps the poster

> **Achieving simplicity on the surface often requires a complex effort**

child for the KISS principle.

Then there is Einstein's caution: "Make everything as simple as possible, but not simpler." Newton's Laws of Motion are simpler than Einstein's Theory of Relativity, but are an approximation that doesn't work when one approaches the speed of light. Avoiding *over-simplification* can be a hard task.

Simplicity on the surface often translates to complexity underneath. "Simple" is hard.

When is it my turn to talk?: Building smart, lean menus

Susan Boyce and Martie Viets

Susan Boyce, Ph.D., principal user interface designer, and Martie Viets, senior user experience engineer, Tellme, a Microsoft company, discuss the tricky problem of building concise menus that don't confuse the user. Susan has 15 years experience in VUI design. She began her career at AT&T Bell Labs where she worked on early speech deployments and also was a member of the pioneering "How May I Help You" team. Prior to joining the Tellme team, Susan also worked for Telelogue designing for Directory Assistance and ran her own UI consulting company. Martie has more than 20 years experience designing telephone and speech interfaces. She began her career at Bell Labs (later Telcordia) and then joined the AT&T Labs team. She's worked on directory assistance, unified messaging, and enterprise applications. Martie has a master's degree in Human Performance Engineering.

"Dear God in heaven…. this isn't what I called about" moaned a customer mid-way through the 55-second preface to her menu choices. Voice interfaces that trumpet all available information to callers leave them overwhelmed, uncertain when to talk, and unclear whether they've even dialed the right number. Though users tolerate—and tend to expect—telephone interfaces to tax their memories and patience, it needn't be that way. Trimming excess information and offering smart, lean menus can produce flexible interfaces that improve not only automation but the user's overall experience.

Skinny it down

A well designed telephone interface ushers callers to the right place and lets them complete tasks quickly. Callers hear just the information they need at the moment. Tips, tutorials and marketing promotions, if included, surface only when they will help the user complete a task, educate her for a later visit, or seed loyalty to the product. Too often, in a well intentioned desire to help the caller, far too much information is loaded into the initial seconds of the phone call.

Take for example this preamble intended to help callers use a brokerage system:

Here are some helpful hints on how to use our speech system. When using our service, significant background noise or side conversation can reduce the system's ability to recognize your request. You can state your request without listening to options in their entirety. Use short sentences or key words to state your commands. For a list of keywords, say More Choices. For useful tips on how to navigate our speech system, say 'brochure' and one will be mailed to you.

Instead, brief introductions and short, flexible menus that tie features together let users get to their task quickly. This approach helps callers navigate easily across tasks, giving them the impression of a fluid, natural sequence.

...brief introductions and short, flexible menus that tie features together let users get to their task quickly.

Nine choices or four? Make mine four...

One example from a redesigned voice interface for a large shipping company's customer-care application produced dramatic results. We consolidated options, trimmed wording, added a Go Back feature for navigation, and modified end menus so callers can move between applications to complete several tasks during a single call.

Original Main Menu	Redesigned Main Menu
Thank you for calling ABC. Please visit us on the web at ABC.com for all ABC service information or select from the following options: ▪ *For ABC tracking information, press 1* ▪ *To schedule an ABC pickup or to receive rate information, press 2* ▪ *For ABC international shipping information, press 3* ▪ *For ABC package time in transit information, press 4* ▪ *To find the ABC store and all other shipping locations and drop boxes, press 5* ▪ *If you received a yellow info notice, press 6* ▪ *To order ABC customer supplies, press 7* ▪ *To hear these options again, press the star key* ▪ *Otherwise, press 0 now.*	*Thank you for calling ABC. What would you like to do? You can say track a package, send a package, shipping information, or order supplies.*

The original opening prompt included:

▪ Nine options
▪ Eleven sentences
▪ 99 words

...and lasted 48 seconds. It invited callers to visit the website before playing menu choices.

Our redesigned opening prompt has:

- Four options
- Three sentences
- 25 words

…and lasts under 10.5 seconds. All of the same functionality is offered.

In the length of time callers used to spend listening to main menu choices, they can now hear the menu, make their selection, track a package, hear the package status, and move on to another task. This improved efficiency has dramatically reduced requests for an operator and has increased customer satisfaction.

Hidden redundancy

To accomplish the dramatic reduction in main menu size, we needed to group the old menu options together into categories. There is nothing new about this. But, as anyone who has tried to do this can attest, it can be extremely tricky to name the main menu choices something that will still have meaning to the caller. If callers cannot accurately recognize the task they wish to perform from the main menu choices, they are likely to go down the wrong path—which has disastrous consequences in most IVR systems.

Several of the tasks were related to sending a package: finding rate information or package transit time, scheduling package pickups, or finding drop-off locations. We wanted to represent all of these tasks with a single menu choice on the main menu. The problem was what to call it? If we called the option "send a package" that might not seem right to the caller who is just shopping for rates. If we call the option "shipping information" this won't seem right to the caller who is ready to schedule a pickup. The solution was to have two options on the main menu that led to the same submenu.

A caller can choose Send a Package to reach the next set of options:

S: Thank you for calling ABC. What would you like to do? You can say track a package, send a package, shipping information, or order supplies.
U: Send a Package.
S: Send a package. If that's not right, say Go back. You can say Schedule a Pickup, Rates, Package Time in Transit, or Find ABC locations.

Or, the caller can choose Shipping Information, and reach the same, trim submenu:

S: Thank you for calling ABC. What would you like to do? You can say track a package, send a package, shipping information, or order supplies

U: Shipping information.

S: Shipping information. If that's not right, say Go back. You can say Schedule a Pickup, Rates, Package Time in Transit, or Find ABC locations.

On the surface, having two identical menus under different options from the main menu seems like poor design. It doesn't appear very efficient when drawn in a call flow. But in fact it has turned out to be a very effective way to eliminate main menu clutter, while still protecting the caller from the perils of getting down the wrong path.

Main Menu Shortcuts and Just in Time Instructions

Another method we used to speed callers to their task was to build in shortcuts. Callers can say the task name at the main menu if they already know what they want to do. The main menu structure of "What would you like to do?" followed by options naturally encourages callers to answer the question with the desired task name.

S: Thank you for calling ABC. What would you like to do? …

U: Schedule a pickup.

S: Schedule a Pickup. If that's not right, say Go back….. Are you shipping air packages that are ready today?

Next, the application determines whether the call can continue to the next automated step by asking "Are you shipping Air packages that are ready today?" This carefully placed question is more effective than reading callers instructions—which they're unlikely to remember—about conforming package standards (the packages must be prepared; ground packages are not eligible for pickup, etc.) before the call continues. Once a customer successfully completes this task without operator assistance, the person is far more likely to attempt it again.

Context sensitive end menus

We structured "end menu" choices to offer callers easy access to tasks they're likely to want next. For example, after hearing the cost to ship a package:

S: Would you like to hear more Rates?

U: No.

S: Is there anything else I can help you with? Say: Transit Time, Schedule a Pickup, Find an ABC Location, or More Choices. (pause) If you're finished, you can hang up.

U: Schedule a pickup.

S: Okay. Schedule a pickup…

Conclusion

Well-designed voice menus offer simple, concise choices, give users flexibility for moving across tasks, and provide just the guidance a caller needs for the task at hand. Reducing the number of menu options, omitting detail that applies to very few callers, and making it easy for callers to complete tasks successfully increases automation and customer satisfaction.

My Big Fat Main Menu: The Case for Strategically Breaking the Rules

Susan L. Hura

Susan L. Hura, PhD, principal, SpeechUsability, discusses how one of the common assumptions in VUI design depends heavily on the application, and discusses a specific case where she was able to challenge that assumption with success. Susan is the founder of SpeechUsability, a consultancy focusing on improving the user experience by incorporating user-centered design practices in speech technology projects. Susan started and managed the usability program at Intervoice as their Head of User Experience, and prior to that worked a member of the human factors team at Lucent Technologies. She held a faculty position at Purdue University in the Department of Audiology and Speech Sciences. Susan holds a doctorate in Linguistics from the University of Texas at Austin.

How many items should there be on a VUI menu? Even for basic questions like this, VUI designers are often forced to derive solutions from research conducted in other fields that we translate to the VUI domain. Often the translation is none too obvious. Case in point: George Miller's 50-year-old research on short-term memory is often portrayed as the golden rule for VUI menu size limits (Miller, G., The magical number seven, plus or minus two: Some limits on our capacity for processing information. *Psychological Review*, 63, 81-97, 1956). Miller's research focused on participants' ability to recall items

> George Miller's 50-year-old research on short-term memory is often portrayed as the golden rule for VUI menu size limit... far from the whole story

presented auditorily—most people could remember between 5 and 9 items, hence the 'magical number 7' title of his famous paper. The limits of auditory short-term memory are real, but they are far from the whole story for how to design a VUI menu. I know this because I have observed many very large menus that are usable, efficient, and well-received by users. How do we explain this seeming inconsistency?

Let's start by defining exactly what users do when they interact with a VUI menu. Users listen to a list of options, consider each one, then select an option they hope will allow them to achieve a goal. This listen-evaluate-decide paradigm is very different from the simple recall task in Miller's studies. The primary purpose of a menu is to enable users to select among options quickly and easily. Therefore, I suggest that we should instead focus on designing a set of menu options that are *descriptive* and *distinct*, rather than worrying about exactly how many options there are.

Descriptive menu options give users the immediate sense of 'I know what that is!' Recognizable terminology in menu-option names enables users to evaluate and discard non-desired options without holding them in memory. Similarly, users will immediately know the difference between *distinct* menu options, which eliminates the need to remember the entire list while making their menu selection. Current short-term memory theorists agree that there are many factors beyond the number of items that affect auditory short term memory, such as word length, frequency and familiarity, and inter-item discriminability (that is, how distinct items on a list are from one another, see Nairne, J., Remembering over the short-term: The case against the standard model. *Annual Review of Psychology*, 53, 53-81, 2002). The importance of descriptiveness is supported by recent research in human-computer interaction showing that meaningful menu-option labeling is more important than information architecture for users (Resnick, M. & Sanchez, J. Effects of organizational scheme and labeling on task performance in product-centered and user-centered web sites, *Human Factors*, 46, 104-117, 2004).

I found that distinctness and discriminability were even more important than usual when designing a recent VUI menu. I started this project with the luxury of an extended period of requirements gathering that included user interviews and analysis of months of call-flow statistics. At the end of my analyses, I was left with a shocking conclusion: I proposed a main menu with 34 options. While I'm not a believer in the magical number 7 theory of menu design, 34 seemed ridiculously high. The reasoning that lead to this menu highlights how you can break the rules of VUI design without betraying the overall goal of creating a usable automated system.

Here are the basic facts of the project:

- The project was to design an internal-facing IVR for a large US corporation whose purpose is to route employee calls to the appropriate live representative.
- There are two distinct groups of users. Field users call multiple times a day. They call to manage every aspect of their job and call almost exclusively from their cell phones. Home-office users call once or twice a year. They typically call from their desks after failing to find an answer on the company intranet. They are calling to resolve benefits or human resource issues.

- Each user group was currently served by its own IVR system with largely overlapping functionality. Current IVRs used very large grammars to support an open-ended initial question ("*To better assist you, please say a word or phrase for the topic you need.*") Many users are frustrated by not knowing what to say, or because of poor recognition performance, and some users employ techniques to fool the system into taking them to a live agent right away. However, perhaps 25% of calls are users successfully and happily using the keyword system.
- The goal of my redesign project was to consolidate the two current systems into one IVR while improving correct call routing. All current functionality (including keywords) was to remain available.
- The client rejected building a statistical language model to better support their open-ended initial question.

I came to a few conclusions immediately. First, I decided we need a menu to serve infrequent callers. These users don't know what to say and they are uncomfortable. This was a troubling decision because I knew that a hierarchical menu wouldn't work. Previous usability testing showed that users were unable to choose the intended menu option for many tasks when presented with a top level choice of 'benefits,' 'policies,' 'payroll,' and 'other.' The picture became even more complex when call statistics showed that the most frequently used functionality was a mix of very high level ("computer assistance") and very specific options ("family and medical leave act"). These facts all pointed me towards a wide, shallow menu.

Because this menu would necessarily be longer, I realized that there must be a way to bypass the menu for frequent callers and those who successfully use keywords today. These facts support a keyword recognition strategy, but recall that improved recognition performance is another goal. To accomplish this, I proposed using a modified version of the current keyword grammar at the main menu in which we significantly prune the number of items, and use weighting based on frequency of use.

A final consequence of using a wide single-layer menu is that we need one submenu for field-sales options because not everyone needs to hear them.

The options on my big fat main menu are grouped into categories; each category ends with an "other" option where a subset of keywords appropriate to that category are recognized. Users are instructed up front that they can barge in with a keyword at any time, and that they can speak menu options at any time. The voice talent speaks quickly here—the entire menu lasts only 61 seconds.

Here is the Main Menu prompt:

> *You can say....Company Directory. Computer Assistance. Say Field Support to hear those options.*
> *For health benefits you can say Unified Healthcare, Rogers, Dental, Healthcare Reimbursement Accounts, Express Refills, or Other Health Benefits.*
>
> *You can also say Payroll, Direct Deposit, W2, Base Compensation, or Other Payroll Questions.*
> *For financial benefits you can say 401K, Trading ID, Stock Options, or Other Financial Benefits.*
> *You can say Human Resources, Peopletools, Employment Verification, Job Posting, Termination, or Other HR Issues.*
> *For company policies, you can say Family and Medical Leave Act, Short Term Disability, Flexible Work Options, Vacation, Retirement, or Other Policies.*
> *Or, you can say Representative if you know you need a person.*

What we have observed in practice is that frequent users barge in and never hear the menu. Infrequent users who have a term in mind for what they need also barge in without hearing much of the menu. It's only when the user does not know what word to say that they bother listening to the menu. And in that situation, having a menu is an asset, not a punishment. There you have it: thirty four options, quick and simple, the users like it, the client likes it. And I broke a lot of rules.

The downfall of many rules of VUI design is that they are presented as generic prescriptions to be followed in all cases. This makes the rules both too broad and too specific simultaneously. The rules are overly specific because they focus on, for example, the exact number of items in the menu rather than on the more important principle of designing within the limits of auditory short-term memory. Rules are overly broad in that they refer to any main menu, with no regard to differences in domain, user groups, frequency of use, or other factors that should influence menu design.

Here, having 34 items on the menu is no burden to users because the options are distinct and discriminable, and users are not forced to listen to the menu if they don't need it. For these users, on this system, the answer to 'How big should a VUI menu be?' is 34 items, no disrespect to Miller's magical number intended.

What's universally available, but rarely used?

Melissa Dougherty

Melissa Dougherty, Principal and Co-Founder, Voice Partners LLC, addresses the potential fallacy in relying on "conventional wisdom" in VUI design. For over fifteen years, Melissa's work has focused on optimizing both the feature set and the user experience to meet the needs of companies and their customers. Prior to founding Voice Partners, she joined Nuance in 1998 to create and lead Voice Interface Services. Before Nuance, Melissa was Managing Director at Greenfield Consulting Group's office in the San Francisco bay area, where she defined product strategies, optimized user interfaces, and developed an end-user knowledge base for companies like Intuit, Sprint PCS, AOL, and Citibank.

Many speech systems have "universal commands"—commands that are available to the caller everywhere, whether they're mentioned in a particular prompt or not. Historically, universal commands have been thought of as a boon to usability. Theoretically, for example, a caller who is confused can always ask for "help," or a caller who didn't hear the prompt because of some sudden interruption can always ask for "repeat." It seems to make sense, but check the data. We find that almost NOBODY uses universal commands *unless they're mentioned in the prompt.* Only "representative" is used consistently whether or not it is explicitly mentioned.

> ...almost NOBODY uses universal commands unless they're mentioned in the prompt

This raises the question: Is the concept of universals viable, or is another notion of "caller support" necessary going forward? Simply put, universals can be a design "crutch": if the designer believes that the caller will ask for help, for the repeat of a prompt, or access to the main menu, they don't have to worry about whether or not the prompt fully matches the condition of the caller. To be successful, designers must instead consider the needs of the caller at every state of the conversation, and design accordingly.

What does the data show?

At Voice Partners, we're often asked to review or redesign an existing speech system. As a first step in each of these projects, we (1) review data from existing systems, and (2) listen to, and tabulate, a minimum of 100 calls per segment. Time and time again, we've noted that universal

117

commands are rarely if ever used, unless they are mentioned specifically in a particular prompt.

Consider "Help." Historically, "help" has been considered a valuable tool for callers. I, myself, remember advocating its use early in this decade. However, we note that across verticals and applications, usage of "help" is between 0-0.3% for an individual dialog state, with most states showing no usage at all. States where "help" is used more often are those where it is directly mentioned in the prompt.

When the caller *does* say "help," they've generally tried at least twice to say something that the system will accept and have experienced multiple levels of error recovery. In a well-designed system, additional information about how to enter information has generally been offered by the error recovery. Because "help" prompts generally contain the same information as the second level error (with a few additions), the "help" prompt is not helpful to the caller: generic detail has been provided by the error recovery, and the reasons for caller's problems are either very specific to their need, or related to being "in the wrong place" in the application.

Other universals, such as "repeat" and "main menu," demonstrate the same pattern. Usage is extremely low, except in states where the command is specifically mentioned.

A few applications offer "go back" as a universal, in the hopes that callers will use the command to navigate when they've made an incorrect choice. We have never seen even one usage of "go back" in the data we've analyzed.[4]

Why don't callers use universals?

The biggest reason that universals aren't used is that callers don't believe they're available, and have no reason to expect them. They're not a part of natural discourse, and some are completely new for the phone. In repeated usability research studies, we've asked callers whether or not they believe it is possible to make requests that are not mentioned in the prompts, and generally, callers feel that they are limited to requests that

---

[4] Note that many clients don't even log the data needed to give them a good sense of how frequently universals are used. Logging is often based on the number of times that the system experiences a certain event, making a request for "main menu," "repeat," or "go back" hard to distinguish from all the other times that a caller "hit" that state. As a result, our knowledge of universals other than "representative" and "help" comes from listening to, and tabulating, literally thousands of calls to our clients systems.

are mentioned by the system. While many callers assume that an over-the-phone system would allow access to a "representative," usability research also tells us that other universals are not expected from an automated system. Caller's past experience with touchtone systems tells them that commands like "help" are not universally available. No one would ever say "help" or "repeat" in a two way conversation. It therefore becomes unnatural for the caller interacting with a speech system.

What's the impact on design?

If a designer assumes universals are available, and will be used by the caller in certain situations, they need not worry about accounting for that situation in their prompting. Given the low usage, however, we need to help a confused caller get to where they need to go (without expecting them to say "main menu"); we need to make sure that those listening to and noting information actually got to write down the whole number (without assuming they automatically know that "repeat" is available); etc. This means that we must be *very specific* about the prompts created for each state, as they must address the caller needs, at that particular moment...*without* adding unnecessary commands to each prompt, increasing the number of options, and thereby increasing cognitive load and negatively impacting caller success.

Let's consider each potential universal command separately:

- *Representative:* Callers will use "representative" to leave the system, especially if we tell them it is available. It is the closest thing we see to a universal. We see two types situations for its use:
  o In any design, there are moments when we can assume, that callers are more likely to need to talk to a representative: when they're told by an insurance company that their claim will not be paid, for example. Mentioning that the caller could choose to be connected to a person at these moments has a positive impact on caller satisfaction, as they believe the system is "trying to help."
  o Because callers expect (and desire) access to "representative" in an automated over-the-phone system, they will occasionally ask for it. As a result, we generally include it in the grammar throughout the application, even if it is not mentioned specifically in prompts.

- *Help:* Our designs generally do not use "help" as a universal. Instead, we support the caller in the following ways, which have been proven superior in usability research:
  o We employ an error strategy that offers additional context-sensitive detail in the second level error prompts. Since this is generally the same

information that designers put in the "help" prompt, the caller hasn't missed anything, and the system is simplified.

o We offer an option which can meet other needs that might be relevant at that time: The need to get to another area of the application, or be helped out with another topic, for example. This type of option is context-sensitive, so that it offers useful information that callers might be looking for at that time.

- *Repeat:* Repeat should also not be used as a universal. In fact, it is only relevant to callers at certain moments in the dialog where information is being read to them, for confirmation or for transcribing for future reference.

  o No-speech time-out prompts account for the need that some callers have to hear prompt information again, as they automatically re-prompt the caller when there was no response to the initial prompt. This is done just as an agent would, at that stage of the conversation (as if the caller said, "What was that account number?"), obviating the need for repeat.

  o Repeat should be offered as an option, in prompts, when callers may need it.

- *Go Back:* Go back is rarely used in applications or by callers. We advise strongly against using "go back," as we consider context-sensitive prompting options to be far superior in terms of helping callers get what they need from an application. Of the few callers that ask for it, even fewer find it useful. Those who find it useful interpret its meaning in the same way as the designer (that is, take me back to the last step in this transaction, NOT take me back to the last transaction I made). However, an equal number will find it a direct route to complete confusion; because their expectations for the "go back" command are different, they could end up in an unexpected spot in the application, losing their sense of how the interaction works.

Taking it further

There are many more strategies we use for optimizing context-sensitive prompting. Usability research has proven that these techniques are the basis for a far superior caller experience, without a false dependence on universal commands that callers probably don't know are there, and will likely not use.

When Your Caller Anticipates *You*—Dialog Support for More Cooperative Conversations

Jeff Foley and Stephen Springer

Jeff Foley, Solutions Marketing Manager, Network Speech, Nuance Communications, and Stephen Springer, Senior Director, User Interface Design, Nuance, discuss how VUI designs can deal with natural, but complex, responses to prompts. Jeff creates and manages marketing and messaging for Nuance's portfolio of telephony products. As an MIT engineer-turned-marketer, Jeff focuses on bridging the gaps between sales, marketing, and development. Steve directs the art and science of user-centered design for speech applications at Nuance, sets global standards for UI best practices, coordinates the implementation of new tools and processes and consults to key Nuance accounts. Steve joined Nuance when the company acquired SpeechWorks, where he contributed to early design of the first SpeakFreely deployment, SpeechWorks' first in-car designs, and collection of unconstrained alphanumeric strings.

What makes for good Voice User Interface (VUI) design in a speech-based self-service application? These days, many organizations judge the performance of their contact centers on both "hard" and "soft" metrics. Hard metrics—such as automation rates, call length, cost per call, and hang-ups/opt-outs—measure how successful the application is at containing callers within the self-service system. Soft metrics—such as first call resolution, customer satisfaction, and caller experience—judge how well the system solves the caller's problem. Of course, a bad VUI design will wreak havoc on both measurements. However, it is surprisingly tempting to make VUI design decisions that boost the hard metrics but concede the soft ones—especially when it comes to confirming caller input.

Fortunately, we are well beyond the "Press or say 1" Dark Ages of speech technology. These days VUI designers can focus on maximizing automation rates *without* sacrificing customer satisfaction by making each call a more natural, more human experience. And that's important—as an analysis of the Nuance Deployment Databank recently showed. By mining data from thousands of deployed applications and millions of transcribed calls, our VUI designers discovered that our speech systems were several times more likely to make a mistake as a result of unexpected input from the caller than from a bona fide misrecognition by the speech engine. In other words, a speech system's performance has become, more than ever before, a function of VUI and grammar design and not dependent only on raw engine recognition accuracy. We often understand the caller's words—but not the intent.

How do these conversations go off track? In many cases, we are a victim of our own success. A conversation with a speech-enabled IVR can deliver a much better caller experience than touchtone button-mashing. However, the dialog flow can be so natural that a caller often inadvertently responds to a prompt as if the caller were talking to an operator. It's not that the caller mistakes the computer for a live person (we've heard numerous examples of callers chiding the system for being automated!), but rather that good design encourages users to react instinctively, and their language instinct tells them to be cooperative and to offer "helpful" information. If the system is not prepared for these responses, the result can be frustrating for a caller. Here's a classic example:

<u>System</u>: "What day will you be arriving?"
<u>Caller</u>: "January thirteenth."
<u>System</u>: "January 30th. Is that right?"
<u>Caller</u>: "No it's not—I said January THIRTEENTH."
<u>System</u>: "I'm sorry, I didn't understand. Is January 30th correct—yes or no?"
<u>Caller</u>: (sighing) "No."
<u>System</u>: "My mistake. What day will you be arriving?"
<u>Caller</u>: "ARRRGH! Just give me an operator." (Starts pounding the 0 key)

The design above demands a two-step process to correct any mistakes: the caller must first simply identify that a correction is necessary, and then separately provide the correction. But this dialog's "did-you-say-A / no-I-said-B" call-and-response is a very natural behavior. Imagine you're at a restaurant and the waiter asks if he has your order right. Would you just say "no" and glare at the waiter until he asked, "Okay, what's wrong?" Typically you would volunteer the correct information or at least point out his mistake. Whether your behavior is driven by impatience or by a more social desire to help him move the dialog forward, it's natural to anticipate where the dialog is going (in this case, to get the corrected order) and to volunteer the information in advance.

This would be similar in a self-service conversation. The caller, trying to be helpful, anticipates the obvious next question: "Well, what did you say then?" Curse those helpful callers! In fact, our Deployment Databank showed that up to 17% of caller corrections are offered in a single step. The reward for their helpfulness? "I'm sorry, I didn't understand—please answer yes or no..." to be followed by further frustration when the system repeats the original question!

As a VUI designer, the easiest way to tackle this problem is to provide explicit instructions in each prompt—for instance, "Is that right? Please say yes or no." Problem solved, right? It's true, we do see an increase in automation rates when prescriptive prompts are forced on the caller. But think about it—when do you hear language like this? Telling people exactly how to talk is, in real life, almost always a sign of exasperation. As in, "Senator, yes or no—do you agree with what was done here?" Or back at the restaurant, where your waiter has returned to the table: "Would you like anything else? Please say yes or no. Whoa, stop—what did I tell you? Yes. Or. No." This kind of prescriptive prompting strategy very sharply establishes a controller/controlled relationship which is in very negative counterpoint to how callers want to be treated by Customer Service departments. It has the unintended side effect of increasing angst and reducing caller satisfaction of the system... not to mention that it still doesn't provide a way around the original problem of forcing the caller to go back, repeat, and reconfirm the previously misrecognized response. Sadly, since automation rates are easier to measure than caller satisfaction, prescriptive prompting is a deceptively simple behavior to adopt as one moves to "tune" a system to the metrics they have at hand.

A more helpful strategy is to add support for what we call "one-step correction" as an inherent part of the dialog flow. If a caller can reject the confirmation by providing a different answer—a technique which callers often use naturally on their own—then the system succeeds on both "hard" and "soft" contact center measurements. It not only decreases costs by improving automation rates and shortening the length of the call—it also delivers a more positive caller experience:

System: "How much money would you like to transfer?"
Caller: "Two hundred fifty dollars."
System: "Transferring $260 from savings to checking. Is that correct?
Caller: "That's two hundred *fifty*... not sixty."
System: "Got it. Transferring $250 from savings to checking. Is that correct?"

Opportunities for one-step correction abound—whether as full or as partial correction:

System: "What's your account number?"
Caller: "Six one seven, five five five, two eight oh three."
System: "617, 555, 2823, is that correct?"
Caller: "No – it's two eight *oh* three."

What's more, adding the capability of one step correction has an interesting ripple effect across the rest of the dialog design, beyond avoiding messy recovery strategies. Since we have a more accessible mechanism for correction, we can relax the confidence thresholds of the recognition engine. This allows us to make better use of implicit verifications to keep the conversation going forward, knowing the caller has an opportunity to step in and correct a mistake:

> System: "From what city are you departing?
> Caller: "Santiago."
> System: "San Diego, California. And when will you be leaving?"
> Caller: "Wait, no – I said Santiago."
> System: "Got it. Santiago, Chile. And when will you be leaving?"

Aha! We've already eliminated at least three unnecessary turns to gather one piece of data. We've skipped the explicit "I think you said San Diego... is that right?" verification. We also no longer need the "My mistake..." re-prompt to ask for the city again. And in between, there's no opportunity for the caller to surprise the system by repeating the correct information as part of their confirmation.

The concept of one step correction has been around for some time, but the programming and QA work required to implement it has made it prohibitively expensive for most applications. Fortunately, applications can take advantage of such advanced techniques by including a conversation manager. HumanTouch applications from Nuance support one step correction natively by using a conversation framework, known as OpenSpeech Dialog. The framework allows the system to go back and correct any piece of information that it's tracking when it recognizes a valid correction:

> System: "So to review, that's a room with a king bed, checking in on Tuesday, January 31st, checking out on Saturday, February 4th, for two adults..."
> Caller: "No, I'd like two queen beds, please."
> System: "Okay. To review again, that's a room with two queen beds, checking in on Tuesday, January 31st..."

No matter how you prompt and coach callers to respond within your pre-defined grammar, they often can't help but try to provide proactive information. Speech technology has advanced to the point where we must not force callers to adapt to the needs of the system. We can adapt the flow of the call to tolerate callers who treat the conversation like... well, like a conversation. While we can't anticipate *every* possible caller response—references to out of domain topics, requests for abilities that are out of scope for the application—we *can* do a better job of handling the most common responses, such as one-step corrections. Design a VUI that is more tolerant of how humans talk to each other, and you'll have a better performing speech application *and* happier callers. Whether you measure "hard" or "soft" metrics, you'll be sure to see positive results.

What'd you say, and other clichés

Bruce Balentine and Bill Meisel

In this column, we vary our format and Bill Meisel conducts a discussion with Bruce Balentine, Executive Vice President and Chief Scientist for **Enterprise Integration Group**, *on a topic of mutual interest. Enterprise Integration Group is a user interface consulting firm. Bruce has been designing user interfaces for more than twenty five years, including fifteen years in the speech industry. Balentine has several publications, including "How to Build a Speech Recognition Application" published by EIG, contributed to TMA Associates' VUI Visions book, and will publish "It's Better to Be a Good Machine than a Bad Person: Speech Recognition and Other Exotic User Interfaces in the Twilight of the Jetsonian Age," this fall through ICMI Press.*

Introduction

We both were skeptical of some of the common phrases used in promoting speech recognition. One of Bill's least favorite, for example, is "speech is the most natural form of communication," leading to overly high expectations of speech recognition systems. In the following dialog, we focus on some of Bruce's least favorites, and Bill plays the straight man.

A dialog on over-used slogans

Balentine: I came across a marketing slogan some years ago: "Speech overcomes the tyranny of the desktop." Such slogans ring well on first peal, but they usually lose their resonance after thoughtful consideration. If the meaning is that the desktop glues us to our chairs, then it is a relative of another common marketing phrase often associated with speech technology: "Any time, any place, from any device."

Meisel: And two shorter cousins might be "pervasive computing" and "mobility."

Balentine: Definitely blood relatives. Apparently the idea is that we are all held back from our computing destiny by the fact that we sometimes sit at a desk to do our work, whereas most human adventure requires that we be out and about. It's dangerous to imply that it would be terribly useful to do things like composing music, designing a building, or running a biological assay "any time, any place, from any device." We need the tyranny of whatever tools we require to do those tasks—reference books, test instruments, large screens—and they may be specialized and lab- or office-bound. Imagine doing any of the tasks while sitting in a restaurant.

Or at an airport. Or while your car is being inspected. Or at an earthquake site. Lost in the desert. Underwater. Crossing a street.

Meisel: So what are we really talking about? What tasks should be targeted for mobility?

Balentine: Maybe white-collar work is not a good fit. Maybe we should devise an alternate list of outdoor work—things that people might do for work or for recreation that requires that a computer be present-at-hand. Candidates might include checking inspection data from a construction site; looking up a license plate number for a stopped car; or guiding a helicopter to a remote mountain rescue site. Now this list is a great argument for a variety of distributed processing, portable devices, telecommunications, GPS gadgets, and field-instrument solutions. Some of them are even good fits for speech recognition.

Meisel: Do these validate the "any place" claim?

Balentine: No, because each has its own unique user interface challenges. They can't be solved with one solution, the implicit claim of the slogans. None of them really fit the "from any device" aspect of the slogan. In fact, all of these tasks probably require specialized devices that must be designed by someone who is very close to the problem. And the "anywhere" doesn't work, either. I would prefer the term, "field computing" to this vague marketing chimera of "anywhere." These solutions are only effective in specific places, for specific reasons, with a specific user interface requirement. And they are all *professional and specialized*.

Meisel: Can you think of a legitimate target audience for "any time, any place, from any device" products?

Balentine: Well, when push comes to shove, we can quickly scramble and come up with:

- Paying a bill;
- Looking up a telephone number or an address;
- Sending and receiving e-mail;
- Getting directions to a destination;
- Checking on a flight departure gate; and,
- Maybe, finding a bathroom.

Do you see what they all have in common? We can already DO all of these things without a speech interface! So why are we pushing so hard to speech-enable them? The reason is embedded in our design culture—

wherein we champion a favorite technology rather than looking with open minds at the comparative value proposition for various media. In other words, the slogans act against us rather than with us.

Meisel: But can't some of these things be done better with a speech interface? And aren't you supposing a lot of resources available anytime, anyplace—for example, a map or Internet access? All of these things— well, other than the bathroom application—can and have been done with a pure speech interface, and can be supported by any wireless phone with a central speech recognition system. The speech interface can not only be a simpler way to interact (e.g., saying a destination address to get directions), it can be a unifying interface, achieving the "anytime, anywhere" objective with a common, easily understood user interface (assuming you designed it).

Balentine: Yes, there can be significant value in a voice interface and in aggregating tasks for appropriate applications. (Otherwise, I might not have a job.) I'm just saying that part of the lack of adoption for speech interfaces results from there often being other ways of doing the same thing, and users are not strongly motivated to learn another way of accomplishing them.

But I'm also saying that a pure speech interface doesn't achieve the anytime, anywhere paradigm even for appropriate tasks if we examine the situation closely. I recently spoke with someone at a conference who was bragging that they had a pure speech interface for an application, no touch-tone. I asked what happened when the environment was too noisy to use speech recognition. He said, "We default to an operator." I asked, "What if there is no operator?" "Well, I guess it won't work," he admitted. I asked innocently: "So you've designed an application that is sure to fail in a common situation?" He changed the subject.

Meisel: Weren't you being a bit hard on the fellow? Certainly, one shouldn't put speech recognition on a pedestal and refuse to augment an application with fallback to touch-tone where possible. But many things one can do with speech recognition, e.g., entering a street address, are almost impossible with touch-tone. And one doesn't want to over-structure an application (creating, in effect, touch-tone hell) just to preserve the ability to easily fall back to touch-tone.

No interface can handle every case, even every common case. It's common for wireless phones not to find a signal in some locations, for example, so that no interfaces work. One might have to walk to a different location to get a signal; one could argue that to use a speech-only interface that one, equivalently, has to walk to a location that isn't too noisy. (Maybe the bathroom application would help here.)

Balentine: But there isn't really any reason not to include touch-tone when one can to reduce the cases where the system fails. A speech-only interface is a particular example of a rigid attitude that enterprise customers regard with incredulity—not only because of the obvious noise problem, but because of the other reasons that we hear callers daily tell us that they *must* have touch-tone: privacy, speed, accuracy, or simply personal preference.

> **...there isn't really any reason not to include touch-tone when one can to reduce the cases where the system fails**

The single most frequent comment among enterprises that have released their first speech application is about DTMF: "If I had it to do all over again, I would not remove touch-tone. Too many users complain that they need or want it at some point during the dialog."

Meisel: So a key part of your message is that advertising speech as an "anytime, anywhere" solution, by any name, can lead us to poor solutions.

Balentine: We must remember when we choose media that we are trying to deliver value to both a buyer and an end user. Cost versus benefit is the first—and often the most—important criterion for design. And, according to that criterion, there's a lot to be said for the occasional pushbutton.

Knowing that from the start helps immensely when it comes to designing a spoken user interface. User interface media must deliver—not just in sexiness, but in the business basics. And simplicity, cost, and reliability are the key design factors. "Any time, any place, from any device"—just like "the tyranny of the desktop"—dissolves in your hands when you really try to grasp it.

But once you've worked through all of your value analysis and found the right fit for speech, then the design falls into place quickly and easily. Not because of slogans but because of effective usability.

Built with Confidence: How Applications Overcome Uncertainty

Patrick Nguyen

Patrick Nguyen, Chief Technical Officer and founder, Voxify, discusses ways to handle possible recognition errors that are adaptive to the specific part of the dialog, including automatic ways of doing so. Patrick has 15 years of experience in corporate strategy, engineering management, and software development. Patrick served as VP of Engineering at Anubis, a data warehouse technology provider, until its acquisition. Patrick was previously Senior Director of Client Solutions for Personify, a provider of e-business analytics software, where he managed professional services and technology integration with application server, CRM, and data warehousing products. Before that, Patrick worked for McKinsey & Co., specializing in high-tech, electronics, and telco. Patrick began his software development career at Australia's Telstra Research Labs. Patrick received an MBA from MIT's Sloan School, and a BSc. in Electrical Engineering from the University of Melbourne.

It is a truism that a speech application cannot serve any purpose unless it recognizes the callers' utterances. Unfortunately, variability in caller accents, speaking styles, background noise, phone line quality, and other audible factors make it impossible for any agent (human or automated) to recognize spoken words with complete accuracy in all instances. Human agents resolve unclear utterances and recover from mistakes by apologizing, repeating questions, or asking for clarification. By modeling these human conversation techniques, a speech application can provide an excellent caller experience even under conditions of imperfect recognition accuracy.

Conversations are fraught with uncertainty. When there is little doubt about caller intent, a conversation should proceed without hesitation. However, when uncertainty arises at any point, the caller should be given the opportunity to clarify, confirm, or correct a previous input. This capability is useful throughout an application, but carries a seemingly high development cost—after *every* regular dialog, additional dialogs (with their associated prompts, grammars and processing logic) must be specially designed and implemented to handle the potential extra interactions with the caller. If these costs could be contained (something that is

> **when uncertainty arises at any point, the caller should be given the opportunity to clarify, confirm, or correct a previous input.**

discussed later), this practice would undoubtedly benefit callers: correct matches would be instantly confirmed while mistakes would be easily corrected, allowing the conversation to flow as efficiently as possible.

The most convenient measurement of uncertainty is the confidence returned by the speech engine about a particular recognition result. This *confidence*—a numerical value from 0.0 to 1.0, where 0 indicates the minimum confidence and 1.0 indicates the maximum confidence—is the speech engine's measure of the likelihood of a correct match. Figure 1 shows the confidence distribution recorded after one prompt in a high-volume speech application in the travel industry. The grammar in this example contained about 3500 distinct semantic values. For this dialog, the most common confidence level was 0.72, with 36% of all utterances having a higher confidence and 64% of utterances having a lower confidence.

To some extent, every speech application adopts a confidence-based approach to deal with uncertainty. For example, No Match handling is a basic technique that rejects a recognition result whose confidence is below a threshold level, and then elicits another caller utterance.

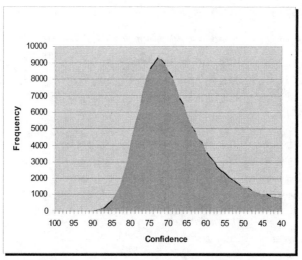

Figure 1 - Sample Confidence Distribution

The categorization of recognition results into No Match and Match events is the simplest treatment of utterances. Sophisticated applications should divide the confidence spectrum (0.0 – 1.0) into more bands to enable finer treatment of utterances. Each band can then be associated with a *resolution technique*—a strategy for clarifying caller intent that includes call flow (for confirming or eliciting information) and grammars (for understanding caller behaviors such as affirmation, correction, and navigation). High-confidence bands would be associated with techniques that require little, if any, subsequent input from the caller to accept a recognition result, while low-confidence bands would be associated with cautious techniques that demand repetition or confirmation.

The Top 5 Resolution Techniques

In Voxify's experience across a number of vertical domains (retail, catalog, direct response, airlines, and travel), five techniques have proven to be generally useful. From the least to the most cautious, these techniques are as follows:

Acknowledgement. This technique assumes the recognition result is correct without further caller interaction. An acknowledgement is optionally played to provide the caller with a cue that his or her response has been heard: *"OK … and what's your last name?"*

Passive Confirmation. This technique combines the playback of the recognized result with the question for the next dialog:

"OK, you'd like to see the Oakland A's. … And what date is the game?"

"OK … On what date do you want to see the Oakland A's?"

The recognized result is implicitly accepted if the caller makes no attempt to change it by, for example, interrupting the next question and uttering a rejection (*"No that's wrong"*), a correction (*"No … I meant the Giants"*), or a navigational command (*"Go back"*).

Extended Passive Confirmation. This technique is similar to passive confirmation in that it plays back the recognized result and then moves on to the next question. The difference is that a Go Back or equivalent hint is inserted to help the caller correct or change the recognized value:

"OK, departing from San Francisco. If that's wrong, say Go Back … And, what's your departure date?"

Active Confirmation. During active confirmation the caller is asked to explicitly confirm the recognition result through a yes-no utterance:

"I think you said 1320 Main Street. Is that correct?"

No Match. This technique rejects the recognition result and asks the caller to repeat his or her entry:

"Please repeat your date of birth. For example June second, nineteen seventy nine"

An Analytical Tuning Methodology

To optimize the caller experience, the confidence levels that separate the resolution strategies should be set using empirical data. Various decision criteria can be used to set the confidence thresholds. For example, one approach is to determine *a priori* the percentage of callers that should experience each technique. An analysis such as that shown in Figure 1 can then be used to select confidence levels so that the percentage allocations are met. For example if the desired distribution of calls is Acknowledgement (10%), Passive (30%), Extended Passive (40%), Active (10%) and No Match (10%), the confidence thresholds would be 0.78, 0.71, 0.54 and 0.40.

Another approach is to set the confidence thresholds based on the *misrecognition rate*. The misrecognition rate is the probability, at a given confidence level, that the caller's utterance was misunderstood by the system. The higher this rate, the more careful the technique should be to minimize the chance of proceeding with incorrect data. For example, it may be decided *a priori* that the Acknowledgement technique is invoked for utterances whose confidence is so high that the misrecognition rate is below 1%, and Passive Confirmation is invoked when the confidence has a misrecognition rate between 1 and 5%, and so on as shown in the following ranges of misrecognition rates:

- 0 - 1%: Acknowledgment
- 1 - 5%: Passive
- 5 - 20%: Extended Passive
- 20 - 40%: Active
- 40+%: No Match.

Normally, a large number of utterances would have to be collected and manually transcribed in order to compute the misrecognition rate. However, a system that initially implements a subset of resolution techniques can be used to estimate the rate automatically. For example, a test system that implements only the Extended Passive Confirmation technique at a particular dialog would be rolled out, and data on usage of the Go Back command at each confidence level would be collected. The likelihood of a Go Back command then effectively becomes an estimate for the misrecognition rate. Figure 2 shows the results of this analysis for the dialog in the travel application presented earlier. As expected, the rate increases as the confidence drops.

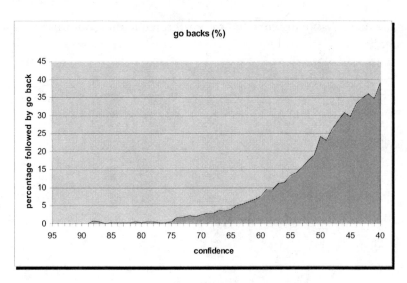

Figure 2 – Misrecognition Rate

The analysis of misrecognition rate needs to be repeated for each dialog in the application. Due to the effect of contextual factors such as dialog location, information type, and prompt design, it cannot be assumed that misrecognition results will be the same from one dialog to the next, even if their grammars are identical.

Using the results in Figure 2, the misrecognition rate exceeds 5% for confidence levels of 0.63 and below. Thus the level of 0.63 should be used as the boundary between the Passive Confirmation and Extended Passive Confirmation techniques in this dialog. By the results in Figure 1, this setting translates into 30% of interactions experiencing the Extended Passive Confirmation or more cautious technique. The majority of interactions (70%) will proceed with minimal impact to call duration and risk of error.

The tuning process itself can also be automated. Continuing the above example, the application could automatically adjust the confidence threshold up or down until the misrecognition rate at the threshold level is at the desired value (this method assumes that the misrecognition rate is a monotonically decreasing function of the confidence level, which is a reasonable assumption).

A Scalable Approach for Implementation

In the industry today, most development tools ignore uncertainty or provide no support for managing it. Designers and developers have to recode and manually tune conversation techniques for every dialog, a practice that is time-consuming, error-prone and un-maintainable. As a result, most speech applications support only the most basic techniques, resulting in rigid strategies that frustrate callers.

Fortunately, new development tools have emerged that embrace uncertainty and support the flexible and adaptive strategies required to deal with imperfect recognition accuracy. Using these tools, call flows and prompts for each technique can be instantiated from standard templates rather than manually defined. Grammars for affirmation, correction, and navigation can be dynamically generated and combined. Thresholds can be continually tuned using analyses of confidence distributions, misrecognition rates, and other behavioral data. Aided by these capabilities, VUI designers can use conversation techniques throughout their applications with little conscious effort.

Although speech applications live with uncertainty, they no longer have to fear it. Modern development approaches allow applications to avoid, detect, and recover from recognition mistakes. Trained with these techniques, automated agents are able to conduct conversations under adverse conditions with the aplomb of their human counterparts.

Beyond State-based VUI Dialogs: Towards Unified Advanced Dialog Design

K. W. 'Bill' Scholz

K. W. 'Bill' Scholz, Ph.D., Speech Technology Consultant, NewSpeech, LLC, discusses the future of dialog design standards and tools, based on his chairing a VoiceXML Forum working group on the subject. Dr. Scholz has over 30 years experience in cognitive science, computer-aided instruction, expert systems, advance software development, and speech / natural language processing. He is a frequent industry speaker and has authored numerous publications. He has consulted extensively for domestic and international organizations in architectural design, speech technologies, knowledge-based systems and integration strategies. Dr. Scholz' primary focus is on integration strategy, software architecture, speech application development methodology, service creation environments, and technology assessments.

For over a decade the majority of voice user interface (VUI) developers have converged on a time-tested approach to dialog design based on discrete dialog states linked by response-contingent transitions. Even as developers bickered over the distinction between user-centered and human-centered design and other guiding principles for building their VUIs, VoiceXML architecture with the Form Interpretation Algorithm (FIA) at its core dominated our conceptualization of any VUI paradigm. However, pressure to evolve ever-greater sophistication in VUI design is stretching the existing paradigm to its limits. Recognition of this fact stimulated the VoiceXML Forum to establish a working group to consider advanced dialog design with the goal of evolving a unified model to lead the VUI profession forward.

> **...pressure to evolve ever-greater sophistication in VUI design is stretching the existing paradigm to its limits**

Initial efforts by the VoiceXML Forum to define a model for advanced dialog design produced a draft of a dialog metalanguage that was described at SpeechTEK in August '06 by Ian Sutherland. Key issues addressed by the dialog metalanguage included:

- VoiceXML applications are in fact *more* than just VoiceXML.
- Pages must incorporate dynamic data and execute back-end functions.

- Design should use Model/View/Controller architecture augmented by extension tags to reference dynamic data and represent common abstractions.
- State Chart XML (SCXML) should be used for the controller component[5].
- The model should be represented in an interface description language that incorporates reusable dialog components.
- Longer-term extensions will be required for more complex dialog structures.

Thus a dialog metalanguage must reach beyond the constraints of VoiceXML and discrete dialog states to manage dynamic data and back-end functions, to articulate a model as distinct from the view and controller, and incorporate extensions for complex dialogs.

In an effort to stimulate thinking in this direction, the Forum decided to sponsor an Advanced Dialog Workshop in which participants were encouraged to share their best ideas on how dialog design must evolve to address the needs of our community. The workshop, organized by the VoiceXML Forum Tools Committee under David Thomson's leadership, took place in February '07 in San Francisco[6]. The workshop agenda focused on three primary topics:

- "States on Steroids" – stretching the limits of state-based dialog design to address future needs;
- Goal Oriented Dialog Design: defining goals, how they can be represented, and how goals and sub-goals can best be managed; and
- Search, Rules, and Stochastic Reasoning: searching across goals and states, representing rules that span states, and prioritizing dialog focus.

The workshop was attended by fourteen senior representatives of industry and academia in the U.S., U.K., France, and Germany[7]. Comments, suggestions, and observations by workshop participants are summarized below.

---

[5] See http://www.w3.org/TR/2005/WD-scxml-20050705/; also SSN, April 2007, p. 35.

[6] Primary workshop organizers included David Thomson, John Tadlock, Ian Sutherland, Jim Larson, and Bill Scholz. Bill Scholz chaired the workshop.

[7] Attendees: David Thomson (SpeechPhone), Jean-Francois Gyss (Orange), Jerry Carter (Nuance), Debbie Dahl (Conversational Technologies), Diego Montefusco (Voice Partners), Srdjan Kovacevic (Acusis), Michael McTear (University of Ulster), Frank Oberle (T-Systems), Phil Shinn (Genesys), Jason D. Williams (AT&T), Silke Witt-Ehsani (TuVox), Bob Wohlsen, John Tadlock (AT&T), Junling Hu (Bosch Corporation)

Extending the Finite State Machine model

Because of its dominance in the minds of many of today's dialog designers, state-based design was considered in depth as the initial topic. SCXML is endorsed by the W3C as a component of VoiceXML Version 3, and is already being used by developers. It is available today from the Jakarta website[8] and provides a state-based execution environment based on Harel State Tables[9]. But as a candidate for advanced dialogs, SCXML would require extensions which would include at least global variables, context-sensitivity, implicit states, a history mechanism, a way to represent multiple simultaneous states, mechanisms to separate state from action, and perhaps even a partially observable Markov process. Detailed consideration of each of these points is beyond the scope of this paper, but one example suggested by Jason Williams suggests what future dialog design may require:

> Consider a travel reservation application with a database containing 10,000 cities. Thus the *a priori* probability of any one city being selected is a hundredth of a per cent: New York=.01%, Los Angeles =.01%, Boston=.01%, San Francisco=.01% and so forth. Then assume a call arrives with a DNIS indicating a west coast caller. Before passing the caller's utterance to the speech recognizer, the city selection probabilities can be adjusted such as New York =.005%, Los Angeles =.3% San Francisco=.5% London=.001%, etc. The recognizer will thus be biased in favor of recognizing a source city on the west coast. Further analysis of this caller's prior travel activities can then prepare for recognizing the destination city by adjusting probabilities such as York =.001%, Los Angeles =.001% San Francisco=.2% London=.001%, Austin=.4%, Boston=.3%, etc. Thus we are, in effect, using expectation of user behavior-based knowledge of past behavior and geographic location to bias recognition. Yet the overall dialog model remains state-based.

We also discussed various problems associated with state-based designs, including the fact that they don't scale well, they require state-definition forms to be filled in arbitrary order, may necessitate recordings of massive numbers of prompts and grammars, tend to be very expensive to develop because of labor intensity, and may require the development of complex

---

[8] See http://jakarta.apache.org/commons/scxml/

[9] See http://en.wikipedia.org/wiki/Statechart#Harel_statechart

mappings of state transitions where transition probabilities are altered by factors such as business model, current application state, user's mental state, etc. Workshop participants considered these issues in some detail and discussed at length the feasibility of adjusting state transition probabilities in a complex state-based dialog using user and domain knowledge as suggested by Jason Williams' example.

Goal-Oriented Dialog Design

Our discussion of Goal-Oriented Dialog Design focused on the definition of goals and sub-goals, and how they would generate actions. We suggested several characteristics of a design's functional and technical goals:

- Goals can be selected based on customer or business drivers
- A goal can combine related tasks
- Goals can be expressed out of order
- Goals need not be confirmed at each step
- Later goals can use information from previous tasks
- States and transitions exist implicitly, but are not defined in advance.

In practice, the process of reaching a goal will require execution of a task; that task may be represented as one or more states in a traditional state-based dialog. The use of goal orientation provides a method for selecting tasks, and customizing the task to achieve a specific purpose. If task selection and customization could be automated, much of the tedium and labor-intensive activity associated with building complex state-based implementations is eliminated. But actual implementation of goal-oriented dialogs still awaits the creation of a suitable infrastructure and development tools. Today's goal-oriented application implementations are highly labor-intensive and require skilled developers. This approach will probably remain as a laboratory curiosity and won't enter the main stream of speech dialog development in the near future.

Search, Rules and Stochastic Reasoning

The third approach to advanced dialog design considered at the workshop centered on the use of rules and stochastic reasoning to guide the progress of a dialog. Rule-based systems attempt to map user actions to application activities using rules. For example, a set of rules may map a user's request for information to specific database queries that are invoked in an attempt to retrieve that information. Subsequent rules may map the retrieved information to explicit responses to the user. Rules, either absolute or stochastic, may be used to select an option from among an

IVR's palette of options, or to alter subsequent responses based on marketing needs, user profile characteristics, or knowledge-base information. In addition, rules can be used to add to a growing body of information in a knowledge base and to implement goal seeking or reinforcement learning, resulting in adaptive systems that grow in sophistication and capability as they incorporate feedback.

Simple rule-based systems have been proven effective in providing advice to call center agents, enabling call centers to provide greater user satisfaction, shorter call duration, fewer misrouted calls, and fewer actual truck rolls. Rule-based systems are also highly effective in implementing diagnostic services, where they implement a decision tree with explicit questions at each node and customer answers defining each branch. Thus, even though the rule set captures the logic central to the diagnostic process, the implementation is based on a dialog state model.

Summary and Next Steps

The workshop considered a broad spectrum of application categories and multiple strategies for their implementation. Clearly there was no single best approach to the specification and development of advanced dialogs. But as we considered sophisticated goal-oriented and rule-based models, we found it convenient to revert to extended state-based architecture as the most practical implementation strategy. We recognized that augmenting traditional state-based dialog by intelligently altering transition probabilities among states and providing global memory of prior activity, could serve as one viable vehicle for implementation of rule-based and goal-oriented systems.

The workshop sponsors have decided to further refine our collective thinking by scheduling the next workshop to take place in New York in August '07. In preparation for that workshop, participants have been asked to contribute use cases intended to stimulate a collective awareness of the activities that must be addressed by an advanced dialog model. Contributors are asked to generate use cases that are as simple as possible, include ambiguity, and illustrate where a state-based model is inadequate or includes transitions that defy enumeration. Our desire is that the careful analysis of a collection of well-conceived use cases will help crystallize our thinking into recommending a feasible approach to the design of advanced dialogs. Once this goal is met, we intend to incorporate all we've learned into refining the specification of a dialog metalanguage. Drafts of the dialog metalanguage will be published by the VoiceXML Forum and contributed ultimately to the W3C Voice

Browser Working Group. Our progress can be monitored at http://voicexml.org/dialogs/.

Applying best practices to VUI design using a vertical market approach

Mike Moore and Jim Milroy

Mike Moore, Business Design Analyst, West Interactive Corporation, and Jim Milroy, Creative Services Manager, West Interactive Corporation, discuss how using vertical market knowledge can improve a VUI design. Mike is a senior member of the West Professional Services team. His responsibilities include speech application analytics, VUI design, usability analysis and testing, and designing integrated applications that incorporate speech/touch-tone and live operator support. In addition to his expertise in call automation, Mike also has 8 years of call center experience and has been involved in virtually every aspect of call center management. As manger of professional services, Jim is responsible for dialog design, persona development, voice recordings, usability testing, and focus groups. Before joining West, he was a professional radio announcer, copywriter, and producer. Jim holds a BS in Mass Communications and Political Science from the University of South Dakota.

Speech recognition technology continues to improve every day. As the voice revolution moves forward, Voice User Interface (VUI) design and development will focus less on the technological restraints of speech and shift to usability and customer satisfaction. In fact, this is already happening today. The purpose of this chapter is to review some of the common best practices for VUI design and demonstrate how to apply them using a vertical market approach.

Common Best Practices

Help callers achieve their objectives quickly and easily

Most automated systems offer a variety of options for callers to select. Unfortunately, many systems include virtually all the available options in one long menu and cause caller confusion in the process. The reality is that two or three call types most likely comprise the bulk of the call volume.

> **The reality is that two or three call types most likely comprise the bulk of the call volume.**

What is the best way to handle this challenge? Here are a couple of recommendations: First, apply the 80/20 rule when designing initial prompts and menus—i.e., cover the primary subjects that a majority of callers will be inquiring about. An easy way to do this, (and simplify the VUI) is to break up the menu structures.

Start by building a main menu that lists out the top two or three most frequently selected options plus an option to hear more choices. From there, drop the remaining callers into submenus. The benefit of this approach is that instead of having every caller listen to a lengthy menu, most callers only hear a couple of items before they find the option they want. This is especially helpful for infrequent callers who do not know they can barge out. An even better way to make the system easy for callers is to place a statistical-model-based (SLM) natural-language speech prompt at the VUI main menu. Natural language allows callers to speak freely in a natural sentence to the automated system. A natural language main menu, integrated with a set of point applications, collapses the menu structure and is more like human-to-human interaction.

Avoid the use of jargon and industry terminology

Usability must be the overriding goal when designing a VUI. Many vertical markets like healthcare and others have their own set of vocabulary or jargon that is specific and known to most in the vertical. However, it is easy to forget that many callers are probably not familiar with the common jargon or terminology of the company they are calling. For example, healthcare patients, unlike providers, *do not know* the industry jargon. Do not force them to learn it by either making patients use the provider application or by using the same industry terms in the patient and provider application. For example, some health insurance payer applications allow patients to call in to check and see if their hospital bill was paid. Payers typically refer to this as "claims status" in their IVRs, yet who's to say that patients like you and I would ever know exactly what "claims status" means—all we know is that we received a bill from the hospital in the mail! Usability studies conducted by our human factors team at West Interactive underscore this point—that patients who want to "check on a bill" are often unsure what the term "claim status" means. Few relate the claim status IVR option to the bill or statement they recently received. This is but one example of many that illustrate how using jargon can be confusing for callers and cause dissatisfaction.

Design prompts that are brief, clear and easy to understand

Avoid adding unnecessary instructions for data that most callers already know how to provide. For example, in the financial services market, if a person is calling to apply for a credit card and the application asks for their date of birth, most callers will intuitively know how to respond. Yet many automated systems think that they must tell callers *how to say their date of birth* using prompts like this: "Please tell me your date of birth. For example, January 2, 1965". Adding instructions makes the prompt longer than it needs to be and can actually confuse callers because they start thinking about *how to say their date of birth* instead of just saying it naturally.

Understanding Vertical Markets

A solid awareness of the vertical markets can help you to design and build caller-centric voice self -service applications. In addition to being well versed in the core best practices of VUI design, such as those cited earlier in this chapter, and in many other industry best practices, dialog designers should also pay close attention to the specific needs of individual vertical markets. Three verticals rapidly adopting speech automation are the media and communication, health insurance, and financial services verticals. The section below includes examples of some key vertical market VUI best practices. The list is not exhaustive, but illustrates the critical role that vertical market knowledge plays in designing a good voice user interface.

Media and Communication Vertical Best Practices

- **Play a proactive outage message.** If a customer is calling from an area where there is currently an outage, there is a high likelihood the caller is experiencing an issue related to the outage. There is a simple way to handle this; provide the caller with the necessary information immediately. When the outage is resolved, place an automated callback to notify the customer.
- **Notify callers of scheduled service appointments.** Use automated callbacks to remind customers of scheduled appointments. Allow callers to reschedule or cancel appointments in the VUI.
- **Allow automated pay-per-view ordering.** A large number of calls in this vertical are from customers wishing to order pay-per-view movies. It is also a good idea to play a message at the start of the call to notify customers of the most popular special events that will drive high call volume. This will help minimize average handle time during call spikes related to the special event by getting the callers who are ordering the event in and out of the system very quickly.

Health Insurance Vertical Best Practices

- Separate call flows for patients and providers. As discussed earlier in this chapter, healthcare patients and providers have different user profiles and characteristics. Patients are infrequent callers who are usually unfamiliar with the jargon of the vertical and they require very helpful dialog. Providers tend to be frequent users of the automated system. They are familiar with the jargon and seek a fast, efficient call flow. For this reason, health payers should have separate call flows for patients and providers.

- Automatically play member eligibility information to providers. When calling to check benefits for a patient, providers also need to know if the patient is covered and for what amount. Automatically offering eligibility information to providers is easier for providers to use than breaking eligibility and benefits information into separate call paths.

- Fax a temporary member ID card. It is helpful for providers to receive a temporary member ID card when they have a patient who either forgot or does not have their member ID card.

- Find a doctor or hospital. Offer functionality to patients that will help them locate a provider in their area.

Financial Services Vertical Best Practices

- **Design prompts that are quick and to-the-point.** This rule applies in every vertical market, including financial services. Today, many credit card companies offer prospective cardholders the ability to call into an automated system to apply for a credit card. Credit card applications require callers to provide a large amount of information, and calls can take a long time to complete. Dialog must be clear so callers understand exactly what they need to provide and also needs to be efficient to keep the call as short as possible.

- **Avoid lengthy sales messages at the beginning of the call.** The fact that a customer called shows that the marketing worked. They do not need to hear the sales pitch again.

- **Set the context and tell caller why they must provide sensitive information.** For example, "The information gathered on this call will be used for credit processing only and will not be used for any other purposes." In addition, some callers may be calling from work or some other public place so *offer touch-tone as an option for confidential data entry prompts.*

- **Capture the most important information first.** The primary goal of the caller is to secure a credit card. Prompt the caller for the information that is required for them to complete their objective first. Do not offer up-sell opportunities or balance transfer functionality until after the caller's initial desired task has been completed.

Concluding Remarks on Dialog Design and Jargon

As discussed in this chapter, formal dialog and the use of jargon may be acceptable if an automated system is communicating with someone who is a member of a defined, known set of callers where there is a set of broadly understood industry terms, e.g., healthcare providers. Contrarily, a casual dialog style, using common, well-known terms is recommended for a system that deals with a relatively open, broad-based group of callers who are not likely to be familiar with industry-specific terms, e.g., healthcare patients and plan members. In our speech industry's effort to continually build outstanding voice user interfaces, we should tailor our dialogs to the specific caller group that will be using the application and incorporate vertical market knowledge and best practices in our design of call flows, dialogs, and applications.

Are You Speaking My Language?

David Ollason

David Ollason, lead program manager, Microsoft Speech Server, Microsoft, discusses issues in moving a speech recognition application from one language to another, usually called "localization." David's career in speech started 16 years ago with British Telecom Research, working on the, then, relatively new area of sub-word unit modeling. He went on to join a start-up, Entropic Cambridge Research Labs, spun out of Cambridge University U.K., and worked on Version 2.0 of HTK (HMM Tool Kit), and later, led the VUI Applications team, building speech interfaces for a variety of applications. The company was acquired by Microsoft in Nov '99, and David continued to lead the Speech Applications team, developing sample applications for MSS 2004 and MS Connect, the auto-attendant currently deployed at Microsoft. He subsequently moved into the role of program manager, responsible for the "authoring experience" in MSS 2007 and VUI design for the Exchange Unified Messaging application, and is now Lead Program Manager for both Authoring and the Platform. He holds several patents in this area, and obtained an MEng in Electrical and Electronic Engineering from Heriot-Watt University, Edinburgh.

With a short chapter such as this, any attempt to cover the full gamut of problems that arise when trying to localize speech applications to other languages would do a serious injustice to the problem as a whole. Issues can start with ensuring that the persona of the application is appropriate for the target culture. One needs to be mindful of plurality differences, and the like, when constructing prompts from concatenated fragments. Grammars may have order-of-information problems and/or reusable portions may need checking for gender and case agreement in the various contexts in which they are used. This is the tip of the iceberg; there are a host of other more subtle issues too.

Rather than deal with these issues directly, I want to fast-forward to a point where all the major issues have been dealt with, and what remains is to put in place some process checks and automated tests that will help in a large-scale localization effort. With many languages in scope, one is unlikely to find localization engineers familiar in both the common task of string translation as well as with the particular problems that pertain to speech application localization.

process checks and automated tests that will help in a large-scale localization effort

Ambiguity Is A Speech Recognition Engine's Enemy

Ultimately, the target grammar content will be created from text produced by the localization process, by localization engineers that don't necessarily understand the representation required by a speech engine. If the resulting text is ambiguous, from the engine's perspective, then Text-Normalization (TN) rules will be employed to resolve the ambiguity. These rules are likely to produce a number of unwanted forms, and potentially not even produce the desired form. For example, how would you write the word version of '55'? Is it, *"fiftyfive"*, *"fifty five"* or *"fifty-five"*? The last form is the one that Microsoft Word® prefers, but may cause TN to insert words like *"dash"*, *"hyphen"* or *"minus"* as alternatives into the grammar, if the speech recognition lexicon needs the two-word representation, *"fifty five"*.

Similarly, one can easily imagine the introduction of punctuation, such as '! . ? ,' etc, which could be rendered literally, or numbers like '15', which could produce the unwanted form, *"one five"*, as well as the expected, *"fifteen"*. Simply stripping punctuation from the text may seem like an easy solution, and indeed it could be appropriate in many cases. However, this may have a negative impact on entries like, *"Mr."*, *"Prof."* etc, which appear as such in the Lexicon.

It is my view that the safest option is to avoid punctuation and ambiguity altogether and to be absolutely explicit as to the desired form in the grammar (e.g. is *"St."* supposed to be *"Street"* or *"Saint"*?). But the localization engineers don't know this, and it is a difficult educational hurdle, so design-time checks that highlight these problems are invaluable.

That Word Isn't In My Dictionary

In a similar vein to the issues described above, one can easily imagine misspelled word entries, or correctly spelled words that are not in the speech recognition lexicon. In both cases the engine will employ Letter-To-Sound (LTS) rules in an attempt to craft pronunciations for the unknown word, and although the algorithm may do a great job in the majority of cases, it is no substitute for hand-crafted pronunciations.

As an example, I have noticed a tendency, particularly among engineers, to spell the word *"forty"* as *"fourty"*, and in the worst case, this could result in a pronunciation akin to "f-hour-ty". Presumably common misspellings exist in other languages too. Obviously, spelling errors can and should be corrected at the source. However, it can also be the case that the speech recognition lexicon is simply deficient in the desired words, or it may be that the words are in some way 'fabricated' - it is all too common for brand names to deliberately trademark the misspelling of words for effect (*Krispy-Kreme*™, for example).

Again, human error and ignorance of the issues requires design-time checks to highlight and allow correction of the inevitable mistakes.

These Sound Similar But Mean Different Things!

The Localization process runs a significant risk of providing a pair, or more, of competing commands in the grammar that are acoustically similar to one another in the target language, but were not so in the source, creating a recognition problem. Suppose US English is the target language and that the task is the commands used in the email portion of the Outlook Voice Access (OVA) application in Microsoft Exchange®. One possibility is that the concept of moving between messages is translated as, *"Forwards" / "Backwards"*, and that the concept of forwarding a message to someone is translated as *"Forward It"* – clearly we now have a likely recognition problem between *"Forwards"* and *"Forward It"*. Solving this problem by expert scrutiny does not scale well to a process involving many languages, and some automated help is therefore required.

Microsoft's Speech Server platform features a technique for automatically identifying acoustically similar, competing phrases. It is achieved by converting the text of a test phrase into a synthetic speech signal and then using the speech recognition process to discover competing words or phrases that are acoustically similar to the test phrase, and then iterating over all test phrases.

If the similar phrases discovered are ones that the application explicitly confirms (e.g. *"fifteen"* and *"fifty"* in the OVA application), then no action need necessarily be taken. However, if, as in the earlier example, the phrases are commands that the application does not confirm, then one of the pair must be changed, in both the prompt and the grammar, to an alternative that is more dissimilar (e.g. use *"next" / "previous"* instead of *"forwards" / "backwards"*).

Are The Left And Right Hands Doing The Same Thing?

It is clearly a good practice to have the same person do both the prompt and the grammar translation, in that order, with a human consistency check between the two. In this context, consistency means that, at minimum, the commands, or example input phrases, spoken in the prompt, are represented verbatim in the grammar and that the semantic meaning returned is correct. As an example, one wants to avoid a scenario whereby the translation of the prompt from US to UK English replaces the phrase, *"Cell phone"* with, *"Mobile phone"*, but the same change is not made to the grammar. Note, this is a poor example in the sense that both phrases should be in the grammar for both languages, but I hope that the point has been made – for applications where the expectation is that the callers will seek guidance from the prompts, the instructions contained therein must parse the grammar faithfully.

The portions of the prompt that might be repeated by the caller should be marked, thus allowing for an automated text-emulation check to be performed against the appropriate grammar at design time, using the marked phrase from the prompt as input.

Text-emulation also returns the semantics recognized for the given utterance. Using this feature, one can test that, for example, the grammar localization process has not resulted in the phrase, *"January second"* being entered into the semantic slot for January 3rd.

Order Sometimes Trumps Naturalness

The focus of this chapter is to discuss automated tests that add real value to a large-scale localization process. However, I would like to finish by saying that, whilst these tests are very powerful and catch many problems early in the process, they do not guarantee high quality localization. Although it is certainly a goal to reduce any learning curve that the localization engineers have to burden, it is inevitable that some amount of education has to be provided for the speech localization scenario, as compared to the regular GUI localization one. For example, localization engineers unfamiliar with speech localization need to understand that they are localizing 'concepts', rather than a 1:1 string translation, and that there need not be a 1:1 correspondence between the number of synonym phrases in the source and target languages.

Additionally, there may be speech recognition requirements that dictate that a certain form must be adopted, regardless of whether this fits well with the VUI goals of the application, and this must be communicated to the localization engineer. For example, consider the commands, *"Find By Name John Smith"* and *"Forward Message to John Smith"*. Both are of the form, 'command' followed by 'name', and must remain that way, even if it is more natural in the target language to have the name come first. If the order is swapped, then potentially many thousands of names are now competing directly with each and every other command that drives the application.

In Summary

The subject of speech application localization is extensive and is very often application-specific. However, there are a number of powerful and generic sanity checks, as well as specific instructions, which can help to significantly reduce the risk and time taken to produce a high-quality localized application.

Spanish, English, or … Spanglish?

José L. Elizondo and Peter Crimmin

José L. Elizondo and Peter Crimmin, Nuance Communications, discuss issues in dealing with English spoken by native Spanish speakers. Jose Elizondo is a manager in Nuance's Professional Services group, where he oversees multilingual VUI design with a particular interest in Spanish speech systems for Hispanics in the United States. He has designed and deployed speech systems in multiple languages for the past 14 years, and has given lectures and conducted workshops on multilingual design in Europe, Asia, South America, and North America. Peter Crimmin is a technical writer for the Enterprise R&D division of Nuance, where he works on speech recognition products. He has a deep interest in Spanish-language literature and culture, and relied on that experience when contributing to this chapter.

Introduction

As automated systems proliferate in the United States, more applications are designed for Hispanics. But there is no standard Spanish that applies to all Spanish-speaking users. Instead, the language varies from group to group and largely depends on their countries of origin. Designers of Spanish systems must take account of language differences when playing audio information to users, and designers of speech systems must take steps to recognize variations in the spoken language.

One approach is to analyze the audience, their countries of origin, and their language preferences. For example, in the United States as a whole, 67% of the Hispanic population is of Mexican origin. In states like California and Texas, that number is as high as 84%, with a small representation of Hispanics from Caribbean countries. In contrast, Florida is 17% Mexican, 41% Cuban, and 18% Puerto Rican. Armed with this knowledge, designers can attempt a relatively neutral Spanish that uses a blend of words and phrases known to most users. This solves many problems, but others remain. Spanish grammar, and especially verb conjugations, will always have a national or regional bias.

After addressing the question of which type of Spanish to use, there are other linguistic features to consider. One that is becoming more important, particularly with the deployment of more powerful applications that encourage callers to speak naturally, is a mixture of English and Spanish known as Spanglish.

What is Spanglish?

Spanglish is defined as a collection of *language contact* phenomena characterized by combinations of Spanish and English. The combinations can happen at the sentence or word level. At the sentence level, Spanglish mixes words from both languages. "Quiero pagar my phone bill, usando mi checking account" combines languages to say "I want to pay my phone bill, using my checking account." At the single word level, Spanglish forms a new Spanish word from English vocabulary. "Parkear" combines the English "to park" with a Spanish inflection and ending. In the United States, this word is commonly used by Hispanics instead of the Spanish word equivalent "estacionar."

Speaking Spanglish is an additional layer on top of regional variations of Spanish. It exists as an additional layer on top of them. Hispanics in the United States use Spanglish words or phrases in their conversations, regardless of the type of Spanish they are speaking. For example, a speaker doesn't use Spanglish instead of Mexican Spanish, but rather in addition to Mexican Spanish. The same is true for any other regional variation and even for the best approximation to a neutral Spanish. (Examples of Spanglish are shown in the box, following page.)

How prevalent is Spanglish?

Attitudes towards Spanglish vary. Some see it as a degradation of language, while others acknowledge its richness and vitality. Many Hispanics in the United States see Spanglish as part of their linguistic identity.

Examples of Mixtures of English and Spanish

Spanish expression exists
"El show" instead of "el espectáculo"

Spanish exists but it's awkward
"Emailear" instead of "enviar un correo electrónico" meaning "to email"

Spanish exists but it's too fancy
"Las appliances" instead of "aparatos electrodomésticos" meaning "appliances"

Spanish doesn't exist
"La Internet", "los ringtones"

False cognates
"Aplicación" instead of "solicitud" meaning "application"
"Forwardear" instead of "redirigir" meaning "to forward"
"Printear" instead of "imprimir" meaning "to print"

Calque (a word or phrase borrowed from another language by

Spanglish is not a fringe phenomenon. Millions speak it everyday, and it transcends the sphere of informal conversation. Spanglish is everywhere: in music, film, websites, magazines, and newspapers. It appears in university curricula and dictionaries.

Spanglish is not a fringe phenomenon. Millions speak it everyday

Spanglish proliferates for various reasons:

- Many Hispanics live in bilingual environments. Switching back and forth from Spanish to English becomes so intuitive that it is no surprise when words and sentences are mixed.
- Memory is a factor. A speaker who can't recall a Spanish word can use English instead.
- Spanglish conveniently expresses ideas using words and phrases from both languages. Speakers enhance communication by choosing the most concise and effective phrases available from either language.

Motives for adopting Spanglish

As companies pay more attention to caller success metrics and caller satisfaction for their Hispanic customers, offering a generic Spanish isn't enough. To reach automation goals, designers must adapt to the way callers speak; and this includes Spanglish. When callers understand the system without effort, and can speak naturally in response to system prompts, transactions are faster and more successful.

The advent of natural language speech recognition accelerates the need to understand the Spanglish phenomenon. A directed dialogue system leads callers into predictable, well-structured responses. "How many people will be traveling?" "Do you want to travel during the day or at night?" But natural language systems encourage spontaneous responses, and for Hispanics this includes the use of Spanglish. "How may I help you?" "Please describe the problem you are having with your email." In some applications, this can mean 15-20% of all responses include Spanglish.

Because Spanglish is widespread and part of the Hispanic identity, major companies have adopted it in their marketing strategies. Budweiser uses Spanglish in Spanish-language posters. One company is called "vueling.com", where the Spanglish "vueling" is a mixture of "vuelo" and "flying". There is even a newspaper named "The Spanglish Times: Our Periódico."

Implications for Speech Recognition

Acknowledging Spanglish as a linguistic reality has implications for Spanish speech systems. The most obvious effect is on recognition accuracy. If a speech system does not accept Spanglish words and phrases, recognition fails, users are forced into repetitions, and system statistics show an increase in agent transfers and disconnected calls.

Spanglish affects recognition accuracy more than variations in accents. If users from different regions pronounce words differently, the system can be prepared. But when users speak Spanglish, the system cannot invent new vocabulary words on its own. Mixing English and Spanish to predict what a caller will say in Spanglish is not a trivial matter. Various approaches are available and include the use of parallel speech grammars, bilingual acoustic models, and user dictionaries with extended phoneme sets. None of these solutions exists as a pre-packaged Spanglish solution and some might not be possible for every recognition engine or platform.

Call Routing systems typically receive many Spanglish responses. When the system asks, in Spanish, "Which department are you trying to reach?" and the user wants the Human Resources department, the answer might be in Spanish ("Al departamento de Recursos Humanos"), English ("Human Resources"), or Spanglish ("Al departamento de Human Resources"). Because these exchanges occur at the beginning of the telephone call, it's crucial to handle the Spanglish utterances appropriately. Moreover, the Spanglish phenomenon isn't limited to the first few interactions in a call flow or to complex recognition contexts. Even Yes/No questions must accept Yes/No answers (and synonyms) in English, Spanish, and Spanglish throughout a phone call.

Spanglish isn't always obvious or predictable. There are some patterns for invented words, but the best Spanglish solutions are data-driven and not simply predictive. Whenever possible, it is advisable to collect data before deploying the system and use the data as the foundation of grammars. But that is not enough. It takes careful analysis and experimentation by speech scientists and caller experience analysts to find the right balance of vocabulary, speech grammar, and acoustic models. The Spanglish solution must be introduced while avoiding extra complexity of the recognition task. Otherwise, the additional grammar permutations and dictionary entries can actually confuse the recognition engine. Tuning is the key to properly understanding Spanglish in the specific context of the speech system. The most effective tuning process tests the system with progressively larger audiences and implements incremental improvements with each iteration.

Implications for Prompts and Presentation

The prerecorded prompts and announcements of the automated system are the voice of the application, and the image of the company. Spanish messages should sound professional and fluent, and use proper Spanish whenever possible.

But Spanish systems, like their English equivalents, are not deployed to teach the language. Their purpose to convey and collect information overrides arguments for linguistic purity. A strict insistence on translating every word to Spanish is incongruous with the business goals of a company, as well as the reality of Spanish callers. Callers might only know the English expressions for "Tax ID" or "mutual funds", and have no knowledge of the Spanish equivalents.

Conclusion

The sparing and careful use of Spanglish can help create a more usable system with a higher automation rate and shorter calls. If a message has the eloquence of Cervantes or Gabriel García Marquez but is difficult to understand, the system is harmed. Designers should use well-placed Spanglish words when it serves to make a question clearer or when it makes the information easier to understand. In the same manner, a system that does not recognize the Spanish that callers actually speak, including Spanglish, fails to serve the goals of the company and its callers.

Spanglish may have started as a fringe phenomenon, and language purists may still disapprove, but it is the linguistic reality of Hispanics in the United States, and it's not going away. The intelligent handling of Spanglish will increase the effectiveness of a Spanish speech system, and will help strengthen relationships by showing respect for the culture heritage of callers.

Choosing the Right Type of Spanish for an Automated System in the United States

José Elizondo and Ilana Rozanes

José Elizondo, Manager, multilingual VUI design, Nuance Communications, and Ilana Rozanes, an independent consultant discuss a real problem that only becomes apparent to those working on the front lines of VUI design. José has been commercially designing and deploying enterprise speech recognition systems for 11 years. José has deployed systems for companies such as Bank of America, United Airlines, and Continental Airlines. José has conducted workshops on Speech User Interface in Japan, England, Canada, Mexico, Venezuela, and the United States. He received degrees in Electrical Engineering and Humanities from MIT, and studied Music Composition and Conducting at Harvard University (and is an accomplished composer, with many works performed publicly.) Ilana has over five years of user interface design and human factors experience, and first entered the speech recognition field in 1999 as a Natural Language Processing Engineer for Lernout & Hauspie (today a part of Nuance). Within Nuance Professional Services, Ilana designed systems in English and Spanish for clients such as Verizon, Bright House Networks, National Grid, and The Hartford, and has led usability studies for companies like AOL and T-Mobile in the US and Telemar in Brazil. Ilana has also led usability studies for speech applications for cell phones. She is fluent in Portuguese, Spanish, and English. Ilana holds a Bachelors degree in Computer Science and a Masters Degree in Linguistics.

What type of Spanish should one use when creating an automated speech-recognition system for Spanish-speakers in the United States? Consider this: if an American company wanted to create an automated system that was both easy to understand and easy to use for English speakers from different countries, including the United Kingdom, India, Australia, and Singapore, the company would carefully select a person with the right qualifications to do the job. Unless that person had experience designing interactive voice response systems (IVRs) for international markets and knew enough about the dialectal and idiomatic variations, he or she might not be able to do a proper job. This condition also applies to the creation of Spanish systems, particularly considering that Spanish has even more dialectal variations than English. Nevertheless, lacking staff specialized in cross-cultural communications, American companies usually take the

Spanish has even more dialectal variations than English

157

following approach: they assign any Spanish speaker to do a translation of the English text, assuming that just because a person speaks the language, he or she is qualified to do the job. Or, they assume that by instructing this person to use "Castilian" or "universal" Spanish, they will be able to accomplish the task successfully and with minimal supervision.

Spanish speakers in the United States come from all over the world. More than 20 countries use Spanish as their official language. Each of these countries has a very unique view on what "standard" or "proper" Spanish is. The average Spanish speaker would not know what to do if asked to speak in "standard" Spanish. At most, he or she might try to avoid regional expressions or colloquialisms, or try to use "proper" Spanish grammar. But even disregarding differences in vocabulary and pronunciation, the definition of what constitutes "proper" grammar (e.g. verb conjugations, use of pronouns, use of prepositions) varies significantly from country to country.

The belief that Castilian is some universal form of Spanish is a misconception. Spanish speakers themselves have conflicting views of what that term means. Strictly speaking, Castilian is one of several languages spoken in Spain. The term is used to differentiate it from other regional languages, such as Catalan or Basque. In other countries, it is used to highlight the common historical origin shared by the many variations of Spanish, rather than to denote a universal dialect that every Spanish speaker knows and understands. Castilian may have been one of the most important seeds for Latin American Spanish in the 1500s. However, the Spanish language evolved in the Americas, incorporating native elements from Nahuatl, Quechua, and other indigenous languages, and during the colonial period, it was also influenced by African dialects. Spanish continues to develop even today, featuring variations that depend on history, geography, education, and even social class. The differences are not simply cosmetic. In some cases, they are an important aspect of the identity of a nation. The term "Castilian" when applied to contemporary Spanish in the Americas, masks the dialectal richness of the Spanish language behind a vague idealization of little practical value.

One simplified way to think of the different types of Spanish spoken in the United States is to categorize them by country of origin. Statistically speaking, Mexican Spanish is the most prevalent. According to the 2000 Census, 64% of Hispanics in the United States are of Mexican origin, followed by 10% of Puerto Rican background and 3% each of Cuban, Salvadorean, and Dominican descent. Based on these numbers alone, it may be tempting to generalize that Mexican Spanish should be used for national IVR deployments in the United States. Moreover, this dialect is very clearly enunciated, has regular rhythm, and tends to be less rushed than other forms of Spanish, making it easier to understand for people from different backgrounds and age groups. However, to determine which dialect of Spanish is best for a particular set of callers, one needs to understand who the callers are, their country of origin and in what area of the United States they live. As an example, it is worth considering that more limited IVR deployments, solely covering areas like New York City or Miami with Hispanics of predominantly Caribbean and Central American backgrounds, can benefit from the use of a more specific dialect.

Choosing Mexican Spanish, or any other dialect for that matter, does not account for the fact that Hispanics in the United States are at different stages in the process of acculturation. For example, first-generation Hispanics may refer to an ATM at a bank as "cajero automático," which is the proper Spanish translation, especially if they were familiar with this service in their countries of origin. But a US-born second or third-generation Hispanic may only use the term "ATM" (as pronounced in English), even when speaking in Spanish. American companies that want to communicate effectively with all their Hispanic customers cannot afford to always use "pure" Spanish, as they risk having customers not understand what is being said. This approach is particularly important for automated systems, because their verbiage has to be understood in real-time by the callers without the benefit of additional re-reading time, explanations by a live agent, or a well-placed hyperlink. If the case justifies it, as in the example above, an American company would be better off using a mixture of English and Spanish, like "cajero automático ATM", which would upset the language purists, but would be more user-friendly and would lead to fewer callers hanging up or being transferred to an agent.

Deciding which verbiage will actually work for an automated speech recognition system, in Spanish or any other language, is not purely a linguistic concern and does not depend solely on who the callers are. Speech technology has come a long way, but it is still safer to use voice commands that are distinct enough so that they are not easily confused by the recognition engine. For instance, "instrucciones" and "transacciones" are a good Spanish translation for English commands like "instructions" and "transactions". However, if they coexist in the same menu of a financial services system, they may create a problem. These words sound practically the same to the recognizer, leading to higher error rates and unnecesary confirmations. Using "actividad de cuenta" as a translation for "transactions," even though it does not exactly mirror the verbiage, would be clearer in meaning and would reduce the frequency of recognition problems.

Another concern that influences the choice of vocabulary is the attempt to use the same programming code for both the English and Spanish branches of an IVR system. For example, when using concatenated prompts to present information, the simple translation of a particular series of English prompt segments may result in a string of words that are incomprehensible to Spanish-speaking callers. Normally, more segments or a different order would be required for such a prompt series to make sense in Spanish. However, a creative programmer with sensitivity to programming issues may be able to produce a creative solution that works, whenever possible, with the same code used for the English system or that requires minimal code changes.

There is no magic answer to the question of which type of Spanish to use in order to get the best results for an automated system. It is not about simply invoking an idealized, yet impractical, concept like Castilian Spanish. Choosing a specific dialect by understanding the demographics of the caller population is a good starting point. But the question goes beyond the realm of linguistics and translation when one considers the technology and programming factors that influence word choices. The right answer comes as an organic result of a simple but thorough requirements analysis process. It is facilitated by having people with the right skills on the project team. And it should be validated by evaluating the results with real customers by, for example, conducting a language survey (or terminology validation study) as part of a usability test. It is mostly when requirements, proper staffing, or usability evaluations are circumvented to save money on the second language that problems arise. In the long run, though, the company will save more money by making sure that its customers can actually use the automated system. What matters is that users understand what the automated system tells them and that the system understands what the users say.

VUI Concepts for Speaker Verification

Judith Markowitz

Dr. Judith Markowitz, president of J. Markowitz, Consultants, discusses how voice biometrics technology (including speaker verification/authentication) impacts user interface design. Dr. Markowitz is a leading analyst in speaker biometrics and thought leader in speech processing. As president of J. Markowitz, Consultants, Judith provides market analysis that includes publication of two monthly monitoring reports on voice biometrics. She chairs the VoiceXML Forum's Speaker Biometrics Committee, is the Forum's liaison to ANSI/INCITS/M1 (biometrics) and ISO/JTC1/SC37 (biometrics), and is an invited expert to the W3C VBWG. She's published extensively and is frequently quoted in business and technical publications. In 2003, Speech Technology Magazine named her one of the top ten leaders in speech.

Increasingly, VUI designers are being asked to add voice biometrics (VB) to their technology portfolios. VB extracts information from a person's speech to determine their identity and/or specific attributes, such as their gender or age. While many VB are useful for VUI designers, most commercial VUI development is directed at speaker verification (SV). SV, which is also called "speaker authentication," examines speech data to determine whether a person is who he/she claims to be.

Consequently, in order to support SV, a VUI must include dialogues for

> **Speaker verification ("speaker authentication") examines speech data to determine whether a person is who he/she claims to be**

- obtaining a claim of identity,
- eliciting speech samples for enrollment and verification, and
- handling exceptions and failures.

The good news for VUI developers is that many of the best practices for and lessons learned from developing VUIs for speech recognition (SR) can and should be applied to development of VUIs for SV.

Obtaining a claim of identity

In order for SV to validate or reject a person's claim of identity the system must have mechanisms for eliciting and capturing an identity claim. The claimed identity needs to be for an authorized user. In order to become an authorized user an individual must enroll in the SV system by providing speech samples and other registration information. The speech samples are analyzed and converted into a voice model (also called a "voiceprint") for the individual. Following enrollment, the voice model is stored in a special database as the individual's "reference model."

During verification, the claim of identity determines whose reference model to retrieve from the database. Knowing the identity of a speaker is even more critical for enrollment because one of the worst things that can happen to an SV system is to enroll an impostor.

Getting the claim of identity often falls to the VUI. Traditionally, that involves DTMF (touchtone) or spoken input of a PIN. A growing number of systems are incorporating other methods, such as the ID of the caller's telephone, into the identity-claim process. A few companies including Porticus, register a user's telephone(s) as part of the enrollment process. SV integrator Authentify includes an outbound call for high-security applications. The image processing capabilities of mobile devices are improving so rapidly that it will soon be possible to reliably use face recognition, fingerprint, and other image-based biometrics as part of the claim of identity or verification. Today, reliable fingerprint identification requires installation of a proprietary sensor.

Eliciting speech samples and handling exceptions

Once the identity claim has been made, the application retrieves the reference model for that identity and the VUI obtains a spoken sample from the claimant. During enrollment, these samples are used to create the reference model for the speaker; for verification, these samples are converted into a voice model that is then compared with the reference model for the claimed identity to determine the validity of the identity claim.

Many operational aspects for creating, storing, and managing voice models are not of direct concern for VUI development, but VUI designers should be aware of them so that the VUI supports those aspects of the system and the security policies involved. One of the best guides covering those aspects is the International Standards Organization's *ISO 19092-2008 Financial services — Biometrics — Security framework* which can be purchased directly from the ISO website. Another good source is *Introduction and Best Practices for Speaker Identification and Verification (2008)* which was developed by the VoiceXML Speaker Biometrics Committee and can be downloaded from my website (http://www.jmarkowitz.com/downloads.html).

Other VUI design considerations include the time required to complete enrollment or verification, the kind of spoken input required, the extent to which the enrollment/verification is done in the foreground, error handling, and the inclusion of speech recognition. Decisions related to many of these issues are governed by the type(s) of SV technology used: (1) text dependent, (2) text prompted, or (3) text independent.

One popular, performance-enhancing trend is to apply more than one of these technologies to the same input using SV algorithms that were developed independently of each other. Some vendors, such as PerSay, have incorporated this approach in their products. Sometimes, application and VUI designers build interfaces to multiple engines and construct fusion algorithms to resolve conflicts. This approach was used in the development of the SV IVR for Australia's CentreLink agency.

Text dependent (TD)

A TD system creates a voice model for the speaker saying one or more passwords. Because they employ both biometric technology and passwords, TD products are often marketed as two-factor security. TD is widely used and supported by most vendors because enrollments and verifications are fast, resemble existing password systems, and are easy to implement. Downsides of TD are that users still must remember a password, the VUI or SV engine should check to make sure that the speaker said the correct password (if not, it may reject a valid identity claim), and passwords can be compromised (requiring resetting).

The VUI's job is to ensure that the SV engine has enough high-quality samples of the claimant saying the password. The nature of the password is generally determined by security or other non-VUI concerns, but the VUI designer needs to weigh in on the acoustic ramifications of those decisions. Most TD engines need passwords that are at least two seconds long and contain enough voiced acoustic data to support good acoustic analysis. Passwords based on a person's name, while convenient for low-security applications, may be problematic when the names are very short and contain few voiced sounds (e.g., Sue Smith).

The VUIs for TD enrollment and verification are usually short and simple. Enrollment involves prompting for three or four samples. When enrollment is done in a single session the VUI normally includes prompt variants (e.g., "After the tone, please say your ID again"), cues that help the speaker to improve the quality of the input (e.g., such as asking the speaker to speak louder), and methods for terminating a session that is not succeeding (called "failure to enroll"). Most verification VUIs simply prompt the claimant to say her/his password. Best practices for handling errors for SR apply to TD SV sessions.

Enrollment that is done over multiple sessions can provide better coverage of voice and channel variability and may even be performed in the background. For example, SV capture may be done in the background when a caller says her/his account number to a telephone-banking IVR. Privacy considerations may, however, require that the VUI inform the caller of such background activity.

Text Independent (TI)

TI systems expect a speaker to use unconstrained, free-flowing speech or, at minimum, say something different every time he/she interacts with the system. Consequently, there is no need to verify that the user has said the right thing and no password to reset. Issues for TI involve ensuring that a sufficient amount of high-quality speech is captured. Some TI systems are able to process as little as 10 seconds of speech for enrollment and 6 seconds for verification, but most still prefer to receive more data.

TI is the only SV technology that can operate entirely in the background. For example, Bank Leumi in Israel uses background SV to verify a caller who is talking with a call-center agent or an SR-enabled IVR. In such applications no VUI for SV may be needed; the data collected in the IVR is simply passed to the SV engine. Such use of background SV raises the specter of privacy and, depending upon the regulatory environment, may be subject to the scrutiny of a government, industry, or corporate privacy watchdog. Privacy issues can be attenuated by storing voice models but not the original analog recordings and by obscuring the link between a reference model and other information about the speaker. If such privacy-enhancements are implemented, it may be useful for the VUI designer to create an automated statement announcing those privacy policies at the start of a call.

Text Prompted (TP)

TP (also called "challenge-response") is a variant of TD that is typically used when there is a strong possibility that fraudsters will try to use a tape recorder to attack the SV system (called "tape attack" or "spoofing"). TP enhances resistance to tape attacks by asking the speaker to repeat randomly-selected items and/or sequences. It would be difficult for a tape recorder to reproduce those random sequences in the time allotted for the user's response. Some vendors have further adapted the TP process to enhance resistance to tape attacks. For example, VoiceTrust periodically adds new items and deletes old ones. A number of other products use SR models, TI, or other means to enable the TP system to include items that the user did not enroll. For example, the VUI for a system that enrolled the sequences 14321, 96979, 38291, and 54321 may prompt for 12345.

The vulnerability of SV to tape recorders is often exaggerated (see Markowitz, J 2007 *Myths and Misunderstandings about Speaker Authentication* at www.jmarkowitz.com/downloads.html). Nevertheless, the threat of tape attacks should be assessed as part of the VUI design, if only to determine whether to include TP either as the primary or a secondary dialogue.

The language in the VUI dialogue can be extremely simple ("Please say..." or "Please repeat") and the items themselves are generally not complex. They often consist of digit sequences, "combination lock" patterns (e.g., 49 52 or 35 21), or a series of individual words phrases. For example, Trade Harbor uses two-word phrases, such as "black coffee." Factors to consider when designing a TP sequence include the user population (e.g., age, technical sophistication, patience) and the level of security involved. Users may become annoyed by long enrollment procedures, so enrollment may be performed over multiple sessions. Verification may also be long, especially if the system presents more than one sequence. As a result, TP is not typically the best choice for the primary dialogue in customer applications. It is better as a secondary dialogue that is used only when the results for TD or TI are inconclusive or a tape attack is suspected. In contrast, criminal offenders in community-release programs may not be bothered by the length of the interactions because using SV is often a privilege that can be revoked.

Completing the Process

During enrollment, once enough samples are captured to create a voice model (or there is a failure-to-enroll situation), the VUI may terminate. As with SR, this may simply involve a thank you or notification that the call is being transferred.

Verification involves comparison of the speech supplied by the claimant with the reference model for the claimed identity. The result is a matching score. A threshold represents a minimum level of correspondence between the two models needed to accept the claim of identity as valid. If the matching score is better than a pre-defined threshold, the identity claim is accepted and the VUI may simply terminate.

Normally, biometric systems use one threshold, but sometimes more than one threshold is needed – especially for systems that expect input from a variety of telephone handsets and channels. The upper threshold reflects high confidence that the caller is who he/she claims to be. The low threshold indicates high confidence that the caller is an impostor. The scores between the two thresholds are often called the "gray area."

The VUI needs to handle matching scores falling in the gray area. That may include a number of options, depending on the needs of the application (e.g., re-prompting in a TD or TP system, use of other authentication, transfer to a human).

The VUI may also need to support variable length authentication, such as that used by the Nuance Verifier which continues to request additional information/input until a specified overall confidence level (for rejection or acceptance of the identity claim) is achieved. The associated re-prompting dialog should be short and the number of requests needs to be constrained before another action is taken (e.g., transfer to a human).

Summary

There are many things to consider when implementing a VUI for SV. This chapter was able to provide no more than an overview of some of them, along with a few useful references.

So you think your VUI is great? Prove it!

James A. Larson

Jim Larson, Speech Applications Consultant, Larson Technical Services, talks about the importance of testing usability. Jim is also Adjunct Professor at Oregon Health and Sciences University and at Portland State University. He is author of the home study guide and reference The VXML Guide http://www.vxmlguide.com/.

You are thinking about buying a 1957 Chevy from Joe's Used Cars. You've kicked the tires, looked under the hood, and checked the odometer. Should you buy it?

Not without a test drive.

You have paid big bucks to well-known consultants to identify a "persona," spell hundreds of words for grammars, write dozens of prompts and error messages, and create pages of dialogs. Will all this result in a world-class VUI?

You can't know without usability testing.

Testing is fundamental to evaluating all user interfaces. Just as you should never buy a '57 Chevy without test driving, you should not buy into a VUI without extensive usability testing.

> **Testing is fundamental to evaluating all user interfaces.**

What to test for?

Some speech applications entertain—they might help us to select which music to hear, summarize sports scores, or even play games with us. Other applications help us perform specific tasks—obtain answers to questions, supply information by filling in an electronic form, or purchase goods and services. Each application should have a single specific goal. We perform tests to determine how well the application helps callers to achieve that goal.

There are several kinds of tests, including the following:

- *Stress tests*—determine how the application works under extreme conditions, such as a catalog sales application during preholiday sales or a time of day application at midnight on New Years Eve. Often these high usage activities are simulated.

- *Function tests*—determine if functions are implemented correctly, e.g., did that electronic payment application really credit the payee account and debit the payer's account?
- *Usability tests*—determine if callers enjoy using the application and/or are able to use the application effectively to achieve specific tasks

While all of these tests are important, this chapter will concentrate on usability testing.

Why is usability testing important?

Usability testing is important to achieve the following objectives:

- *Set common expectations*—both the developer and the customer use the results of usability tests to determine when development is complete and the customer accepts the application. The developer implements the VUI so that it satisfies the usability tests, and customers have a reasonable assurance that the VUI will provide the stated benefits.
- *Measure improvement*—developers use usability tests as yardsticks to measure how well the application performs. Testing not only indicates positive or negative improvement, but also quantifies the amount of improvement. The developer optimizes the application to meet or surpass the usability tests. These tests determine when iterative testing and fine-tuning is complete.
- *Enable comparison*—similar applications from different vendors can be compared using the same usability test. For example, if callers of Application A consistently perform more efficiently than callers of Application B, then Application A can be said to be "better" than Application B.

Everyone agrees that usability testing is vital during application development. Usability testing is also important during requirements collection (in the form of focus groups), during implementation (in the form of "Wizard of Oz" experiments), and after deployment (to determine the need for tuning and maintenance).

Developers must define metrics to measure the degree to which the application succeeds in helping callers achieve their goals. A group of VUI designers attended a workshop at SpeechTEK 2005 and identified ten usability metrics for measuring effective voice user interfaces [see "Ten Criteria for Measuring Effective Voice User Interfaces," *Speech Technology Magazine*, November/December 2005]. These metrics fall into two general categories—preference and performance.

Preference tests measure the caller's likes and dislikes

Preference metrics are subjective. Measure these metrics by asking callers questions after they use the application. Some examples of preference metrics are:

1. *Caller Satisfaction*—the degree to which the VUI meets callers' expectations;
2. *Ease of Use*—callers' perceptions of using the application;
3. *Quality of Output*—callers' subjective ratings of voice intelligibility or voice quality; and
4. *Perceived First-Call Resolution Rate*—perceived successful completion rate on the first call, including both VUI and possible interaction with a human agent.

The above preference tests are generic. They apply to almost every speech application. Developers should refine the questions for the callers to be more specific and relevant to the goal of the application. For example, if the application is an online payment system, Preference metric 4 might be restated as "Measure the caller's success or failure to complete the payment for each of a dozen companies."

Preference testing is accomplished by collecting preference scores from callers after they test the system. After testing the application, developers conduct interviews with callers, who score the various preference criteria. Or callers are asked to enter preference scores onto a paper questionnaire, a Web page, or a verbal VoiceXML form. There are independent companies whose business is to conduct usability tests.

Performance metrics measures how well the application performs for the caller

Performance metrics are objective. They measure how well the application performs. Several callers test the application, with aggregate scores calculated for each metric. The aggregate scores reflect the overall measure of how the application performs. Here are some example preference metrics:

1. *Time-to-task*—the time it takes from answering the call to the time the caller starts performing the desired task;
2. *Task rate*—the percentage of calls that trigger a specific task start-point or end-point;
3. *Task completion time*—the time to complete a specific task;
4. *Correct transfers*—the number of calls successfully transferred to the correct party;

5. *Abandonment rate*—the percentage of callers who hang up before carrying out a specific task; and

6. *Containment rate*—the percentage of calls not transferred to human agents.

The above preference tests are also generic. Developer and speech application specialists should refine the questions to be more specific and relevant to the goal of the application. For example, if the application is an online payment system, time-to-task (metric 1) might be restated as "Measure the time between when the caller begins a payment transaction to the time it is completed successfully."

Three phases are usually necessary to conduct preference tests:

1. *Instrument the application*—Developers mark events in the application which should be recorded onto a log file. Developers select events based on their understanding of the application goals.

2. *Execute the application*—Test specialists create scenarios instructing subjects to perform specific tasks without being told how to perform each task. Multiple subjects from representative caller populations follow the instructions; relevant events are captured onto a log file.

3. *Report analysis*—An automated report generator parses and filters events listed in the log file to create analytic information to be reviewed by the user interface designers and application implementers.

Summary

Developers must clearly understand the application goal, before developing metrics to measure how well the goal is achieved. Preference (subjective) and performance (objective) tests are performed to determine the degree to which the application goal is achieved. Callers answer questions in post-test surveys to supply preference information. Report generators calculate performance measures from log files containing events generated when callers performed the prescribed tasks.

Both preference and performance tests are important. Preference tests indicate how well callers enjoyed using the application, while performance tests indicate how effective the application helps the caller perform tasks.

Would you buy a '57 Chevy without test driving it?

No!

Should you accept a speech application that does not pass preference and performance tests?

Definitely not!

Emotion is the Key to Intelligent Design: Making VUIs Natural and Expressive

Sheyla Militello

Sheyla Militello, in Marketing & Business Development at Loquendo, argues the sometimes-controversial point that Voice User Interfaces (VUIs) should react more like a person would. Militello graduated in Psychology in 1993 and got her Master in Ergonomics at the Polytechnic of Turin. She worked for CSELT (later Telecom Italia's research and development labs) from 1993, where she was in charge of User Interface Design and Usability assessment in the Voice Services and Applications department. She joined the Voice Solution and Professional Services department at Loquendo in 2001. She has over fifteen years' experience in the design and assessment of vocal interfaces and multimodal human computer interfaces.

The necessity of at least partial automation of customer support via Customer Relationshop Management applications (mostly due to economic reasons), is anything but new. There are a huge number of companies that have considerable costs and resources dedicated to call/contact centers and that depend on them for a significant part of their business and support.

There is usually some conflict between the needs of people calling in, and companies that provide the service (see, e.g., *Voice Enabling Web Applications: VoiceXML and Beyond*, Kenneth R. Abbott, 2001, pp. 88-104). Users calling in want to have their needs courteously and quickly met. From the end-user's point of view, it is much appreciated when having a well-trained person on the organization's front-end. From the provider's perspective, well-trained customer service representatives are expensive. At the same time, providers want the call centre experience to please customers in a cost-effective manner. Any conclusion? Users and providers do not use the same criteria when measuring VUI effectiveness and usefulness. The user's criteria are, for example, "Can I get the information or perform the transaction I want?"; "Is the result worth my effort to get it?"; "Do I feel like I'm receiving a valuable service?". While the provider's criteria are: "Does it reduce the load on customer service agents?"; "Are users satisfied with the experience?"; "Does it increase the number of users I connect with?".

> **Users and providers do not use the same criteria when measuring VUI effectiveness and usefulness.**

172

By these criteria, a VUI that gracefully and elegantly ends up routing most calls to a human operator is not meeting providers' needs. On the other hand, a VUI that never routes a call to a human operator is not meeting the needs of some end-users. Achieving a balance between the sometimes-conflicting requirements of end-users and providers is part of the design process. Skewing the balance too heavily to the provider's side often results in a VUI that is overly comprehensive, loaded with options, and frustrating to use.

But is it really possible to automate a call center without compromising the user's needs? One significant finding emerged from a poll carried out by a market research company for the Market Validation "Vocal Browsing" project. Two interaction modes were compared - DTMF vs speech recognition - in an automated call center architecture. One system comprised several DTMF menus many levels deep, and the other a VUI that routed the call automatically towards a category by recognizing the caller's requirements, but then continued with sub-dialogues and requests for confirmations. Findings demonstrated that more than two-thirds of consumers preferred vocal interaction. The reason given was that, even though the system was not completely free of errors, the response feedback received was felt to be "almost human."

The answer, therefore, is YES—*provided that* intelligent interfaces, speech recognition and other technological prerequisites are not automated to such an extent that they are NOT able to deal with the many human and emotional subtleties that a system of that type should be able to manage.

Assuming that all the technologies required are fully mature, therefore, which is the best type of interface for such a system? Much emphasis is being put on avatars and/or digital assistants. Are such solutions consistent with expressivity? Some argue that if an avatar is not able to communicate or understand emotions then it is NOT credible. Any software that has the "human face of an avatar" should ideally also be able to perceive or communicate to some extent the emotions of the person who is speaking, and the technology must therefore make inroads in this direction if virtual assistants are to have a lasting and credible role in human-machine interfaces.

The growing interest in VUI design is inspired in large part by the goal of supporting more natural human-computer interaction. However, this goal becomes particularly challenging when voice interfaces function in difficult user conditions. Applications that range from systems for education and training, to mobile usage in noisy environments, to transactions in natural living environments, are usually very complex, since they need to rely on effective natural language understanding, robust dialogue processing, and context-appropriate speech generation. Users' perceptions of the overall performance of VUIs can be greatly influenced by weakness in each of these different modules.

That effective error-handling techniques are crucial in developing VUIs is beyond doubt (see, for example, Deborah Dahl, "Practical Spoken Dialog System," *Text Speech and Language Technology*, Vol. 26, pp. 41-63, 2005); however, some researchers have recently introduced the importance of the concept of the "listening behavior" of a conversational agent. Actually, while human-to-human listening behavior relies on a complex interplay of gestures and speech feedback, most conversational agents stay silent and passive while humans are speaking to them, such that the lack of a continuous visual and sound feedback may be unsettling. Actually, providing linguistic feedback may greatly reduce that uncomfortable sensation. Linguistic feedbacks are mechanisms which convey understanding, contact, and reactions to the communicative content. The ability of the component technologies (speech synthesis, dialogue modules, and the face of the agent) to provide linguistic and visual feedback is an emerging requirement of VUI design, since a more complete feedback may compensate for recognition errors and increase the usability of the voice interface.

A further emerging requirement deals with the appropriateness of the feedback with respect to the emotional content of the communication. While the recognition of emotions in speech is still a long-term research issue, text-to-speech technologies are beginning to include a variety of expressive cues that can be used in designing the interface when the basic emotional state can be detected on the basis of context analysis. A variety of VUIs and embodied conversational agents can, in fact, be developed for certain given types of users, for example for elderly people. In the latter case, if a system is being developed for facilitating the usage of the web by the elderly, it would be important that the interface be able to express encouragement—to persist despite of initial difficulties, for example. "Encouragement" is not a basic emotion, but an expressive feature that implies both awareness of the difficulty encountered, i.e. context-awareness, and appropriate voice characteristics, such as the right speaking rate, the appropriate intonation, and so on.

This and other related issues are currently being studied within the framework of European research projects, such as Companions (www.companions.org). The problem of appropriateness of feedback is also present when advanced mobile service systems provide critical multimodal communication support for emergency teams during rescue operations, such as the application scenarios studied in the SHARE project during the past few years (2004-2007). Within the duration of the project the consortium developed a very advanced prototype of a mobile information and communication system for large-scale rescue operations. This prototype (SHARE system) is a multi-user system for on-site cooperation that supports the work of fire fighting organizations in the field. The rescue teams can benefit from mobile and bi-directional communication based on 3G (UMTS) infrastructure and mobile WLAN networks located on-site during the emergency operation; the SHARE system allows exchange of structured multimodal information resources, including audio, video, text, graphic, and location information. Vocal technologies have a fundamental role in multimodal interaction when hands-free interaction is necessary. In particular, the speech synthesis has to be robust enough to work under extreme conditions, where intelligibility has to be guaranteed. This requirement can be met only if the speech generation technology presents high pronunciation accuracy and appropriate speech fluency.

So, then, what about emotional content in VUIs for *The Design of Everyday Things*? "Emotion" is today one of the key words in the world of design. In the same approach as emotional design ([see Donald Norman, "Emotional Design; why we love (or hate) everyday things", 2004] one recognizes that previous conceptions of the interfaces and objects of everyday use, all geared towards functionality and usability, were limited and limiting: that is to say, we can not ignore the pleasure, or otherwise, that we get from objects that we use every day. That which each of us is, is also determined by the objects that we use: we choose them, we appreciate them not only for the function they perform for us, but also for the sensations that they give us. If pleasing objects perform tasks better, why shouldn't more agreeable artificial agents also perform better?

If the user experience of speech-enabled systems will increasingly take place between the telephone and the TV, leaving space for more complex person-system relationships, the realization of an assistant that can become a personal companion is not so far into the future. In the case of the voice command Media Center (SSN, April 2007, p. 20), the choice has been made to equip the TTS with a portfolio of customizable voices, enriched with commonly used expressive phrases in order to achieve a consistent profile of the assistant's personality that can guide the user in managing their files in various media.

In conclusion, therefore, any use of text-to-speech should pay close attention to the expressivity and naturalness of the synthetic voice, and these expressive elements should, in turn, reflect the emotional state of the user as far as possible. An inventory of expressive cues as discourse markers can play an important role in such a process, improving the naturalness and expressivity of generated speech. It is important to aim for natural-sounding speech in artificial agents, but it is equally important not to overdo this at the expense of intelligibility and fluency, which in some contexts is of greater importance than expressivity.

Artificial personal assistants will only play a useful role in the relationship between human and machine if the interface is built around the linguistic and emotional content of the human-assistant dialogue, taking into account such factors as politeness, humour, mood, etc.

The Forgotten Component—The Impact of Human Capabilities on VUI Design

Deborah A. Dahl

Deborah Dahl, Principal, Conversational Technologies, discusses how understanding the capabilities of a user of a dialog system can inform the VUI design. Conversational Technologies provides speech and language technology consulting services. Dr. Deborah Dahl is a consultant in speech, natural language, and multimodal technologies and their applications. She is the Principal at Conversational Technologies, which provides reports, analyses and design services that assist her clients in understanding, applying and transitioning these technologies from the laboratory to the marketplace. In addition to her technical work, Dr. Dahl is active in speech and multimodal standards. She is the Chair of the World Wide Web Consortium's Multimodal Interaction Working Group, serves as one of the editors of the EMMA (Extensible MultiModal Annotation) specification and is also active in the Voice Browser Working Group. Dr. Dahl received her Ph.D. in linguistics from the University of Minnesota in 1984. Dr. Dahl has over twenty-five years of experience and has published over fifty technical papers, including the book, Practical Spoken Dialog Systems. She also writes a bimonthly column on standards for Speech Technology Magazine.

Consider a typical system diagram for a spoken dialog application. Along with boxes representing the technical components—the speech recognizer, the VoiceXML interpreter, the web server, the database etc.—we usually see, tucked off in a corner at the edge of the system, a black box labeled "user." While developing good speech applications obviously requires understanding the capabilities of the technical components of the system, the importance of understanding the capabilities of the user is less obvious. However, the user is by far the most complex and hard-to-control component of the system, with limitations and abilities that are enormously different from those of a computer. This chapter gives some thoughts on how systematically studying the properties of the user can not only provide insight into the basis of accepted design principles but also can suggest new capabilities for future systems.

In the last ten years, as speech applications have become more and more widely deployed, the industry has learned a tremendous amount about VUI design. However, much of this knowledge has been acquired by trial and error. Very little work has started from the perspective of the

177

users, their characteristics and capabilities, and the resulting consequences for VUI design. I believe that we will be able to greatly improve human-computer dialogs if we step back and think through the properties of the user as a part of the system.

Very little work has started from the perspective of the users, their characteristics and capabilities, and the resulting consequences for VUI design.

Consider the following capabilities of humans.

1. Social

One exception to the lack of attention to the user's capabilities is the work of Clifford Nass and his colleagues [references 1.2, end of chapter], who have shown how humans bring social responses to their interactions with computers. For example, even though people may know that they're talking to a machine, they nevertheless try to avoid acting in a way that would hurt its feelings, a response carried over from humans' use of language with other humans. This work can inspire us to gain insight into human-computer interaction from not only the social aspects of language use, but from the cognitive and communicative aspects as well.

2. Working Memory

Humans can accommodate only a few items in their working memories. Some of the consequences of this fact for VUI design are well-known and are frequently mentioned in the literature [reference 3, p. 31]. Memory limitations are the basis for an important design principle in such tasks as constructing sets of choices. A set of choices that has to be kept active in the users' minds until they make a choice has to be kept very small. Memory limitations are also important to consider when information is spoken back to a user. Users have difficulty remembering complex information provided by voice. Good designs compensate for this by giving users time to write something down or by offering to send a text message with the information. There are also differences between people in their working memory capacities, for example, older users tend to have less working memory capacity than younger users [4], so applications designed for an older demographic may need even shorter menus.

3. Attention

Another recognized VUI design principle (for example, see the excellent chapter on minimizing cognitive load in [5]) is the fact that humans do not divide their attention well among multiple tasks. Psychological studies have shown that people have a limited amount of attention and

178

that attempting to try to do two or more things at once degrades their performance on everything they're trying to do (see [6] for a detailed analysis of studies of human attention). More complex tasks require more attention. This has several implications for VUI design:

- Understanding complex instructions consumes attention. For this reason, instructions should be as simple as possible.
- Avoid asking the user to do several things at once—for example the user should not be asked to try to understand a complex VUI while simultaneously providing information about the task.
- Attention often wanders if people are presented with something uninteresting or apparently unrelated to their goals. If people are thinking about something else, they will have little ability to follow the system's instructions. This is the source of the common observation that long introductory messages and unasked-for help will be ignored.
- Distracting aspects of the VUI (for example, an over-the-top persona or non-semantic variation in prompts) also consume attention and will make it more difficult for the user to attend to the prompts.
- It's incorrect to assume that all of a user's attention is available for performing the task of interacting with the VUI. The user might well be trying to perform another task while interacting with a system—for example, driving or reading their email—or they might be interrupted by a co-worker, family member, or pet. Consequently, applications should be designed so that momentary distraction or inattention on the user's part isn't fatal to the interaction.

4. Meta-dialog

The human characteristics of memory and attention we've described so far place limitations on VUI designs. However, we can also look at human abilities that go beyond what most computer dialogs are currently capable of. Let's consider some human capabilities and how they might provide the basis for new computer-human dialog capabilities. One is meta-dialog. Meta-dialog is dialog *about* the dialog. Humans can freely step back from a dialog about a specific task, talk about it, and then return to the dialog at the point where they left off. For example, a user might make a request for the system to speak more slowly or more loudly, to pause the dialog while they attend to something else, or to repeat an utterance that wasn't clearly heard or understood. Current VUIs themselves sometimes initiate meta-dialog ("Sorry, I didn't understand that"), but most designs handle only a few aspects of the user's meta-dialog. Consider designing systems to accept utterances like "just a second," "sorry, could you speak up," and "explain that again," act on them, and then resume the dialog at the correct place.

More sophisticated meta-dialogs might include references to the history of the dialog. Users notice that they've returned to the same state that they were in five minutes ago. Today's VUI systems will behave differently if the speech recognizer hears the same thing three times in a row, but shouldn't they also notice if the user is in the same state, giving the same responses, as five minutes ago?

Summing up

Like the human social characteristics studied by Nass and his colleagues, human cognitive and communicative capabilities underlie many commonly accepted VUI design principles. However, these capabilities haven't been systematically studied for their implications for VUI design. Reviewing these capabilities will reveal additional techniques, for example, those based on human meta-dialog capabilities, which can be used to make future VUI designs much more natural.

References

[1] B. Reeves and C. Nass, The media equation: How people treat computers, television, and new media like real people and places. New York: Cambridge University Press, 1996.

[2] C. Nass and S. Brave, Wired for speech: How voice activates and advances the human-computer relationship. Cambridge, MA: MIT Press, 2005.

[3] J. A. Larson, VoiceXML: Introduction to developing speech applications. Upper Saddle River New Jersey: Prentice Hall, 2002.

[4] Craik, F.I.M., Memory changes in normal aging. *Current Directions in Psychological Science, 3,* 155-158, 1994.

[5] M. H. Cohen, J. P. Giangola, and J. Balogh, Voice User Interface Design. Boston, MA: Addison-Wesley Publishing Company, 2004.

[6] H. E. Pashler, The Psychology of Attention. Cambridge, MA: MIT Press, 1998.

VUI versus Contact Center Design

Ron Owens and Fran McTernan

Ron Owens, account executive for Aria Solutions, and Frances McTernan, Sr. Development Manager, Nortel Communications Enabled Business Solutions, Nortel, talk about focusing on the end-to-end user experience, which goes beyond "persona" and other classical aspects of VUI design. Previously, Owens as Director of Multimedia Professional Services at Nortel was responsible for managing the design, development and project management of IVR, contact center, and high-capacity messaging solutions in the Americas. He has over 18 years experience in technology, banking, and voice automation, and prior to joining Nortel, worked at companies such as Intervoice, EDS, Bank of America, and most recently First Data, where he was Vice President of Product Strategy. Ron holds an MBA from Old Dominion University in Norfolk, Virginia. Fran McTernan has worked in the IVR field for over 20 years, the last eight of which she's spent specializing in designing and deploying speech recognition applications. She leads the team of speech specialists in Nortel's MultiMedia Applications Professional Services and has been involved with the successful deployment of speech apps for utilities, telcos, and railroad companies, among others.

The 1968 movie classic *"2001: A Space Odyssey"* helped define the role of VUI (Voice User Interface) design nearly 30 years before the first self-service speech application was deployed in the early 1990's. And to be clear, it didn't make the job an easy one. That movie set an expectation and a perception that humans and computers can interact naturally and easily. We know that even today, talking to a computer is not as easy as we humans expect it to be. When a computer talks to us and invites us to talk back, we expect HAL (the movie supercomputer with feelings) to be on the other end of the line.

When the first commercial self-service speech application was deployed, continuous speech recognition systems performing at acceptable accuracy levels were finally possible. It was clear to the first speech developers, however, that the interface between the caller and the system would need to be carefully constructed. Speech recognition technology was still new to the general population when it was first introduced into the self-service arena. It was essential that the voice user interface between the caller and the system be as clear as possible to maximize recognition performance. It was also recognized early on that introducing speech as the medium to interact with computers would spark some very interesting reactions from the general population. People didn't expect to talk to computers but once we were asked to talk rather than press numbers on the keypad, well, the earliest VUI designers recognized the possibility of opening up the proverbial "can of worms." Why? Because using the most natural and basic form of communication between human beings to interact with a computer would automatically ignite expectations far, far beyond what was really possible. The experience of talking to a computer for the first time is probably very much like being in a foreign airport and hearing someone speaking your language. There is a relief of being able to communicate, and the very real expectation of being understood.

While we like to believe that HAL is real, each response from a caller requires a recognition grammar with a finite, limited set of possible answers. If the application needed to hear a "yes" or "no" reply, it was, and still is, critical to keep the caller focused and engaged to get the desired response. The success of speech applications depended upon the clarity of the user interface, especially when the technology itself was in its early stages. The role of "VUI Designer" blossomed from the desire and the need to maximize overall speech application performance, not necessarily as a move towards the ultimate goal of "usability," or achieving the best possible balance between automation and customer satisfaction.

VUI design emerged as a critical phase of deploying successful speech applications, and the benefits of actually developing the user interface, rather than letting it happen by default, were coming into focus. The necessity of planning and designing the voice user interface gave us the opportunity to even consider customer satisfaction and the possibility of a self-service application that is all at once usable, successful, and *liked* by callers. It's clear that the VUI was directly responsible for bringing us closer to achieving the positive balance of automation and customer satisfaction.

The idea of actually manipulating the interface to maximize success was so exciting and seemed so fresh-and-new, however, that the industry took the craft to an extreme and essentially narrowed the focus only to speech recognition applications. Rather than viewing the entire customer experience from call inception to ultimate resolution, whether through automation or via agent assistance, designers started to venture into the "persona waters." The concept of attaching a personality to the VUI itself was considered the natural progression to making voice user interfaces mimic human interactions as closely as possible. Speech designers started to hyper-focus on attributes on the VUI's persona—names, descriptions, hobbies, education, and even in some cases, the persona's lineage and family history.

Unfortunately, while the industry focused on *who* the system was, we began to lose sight of the prize: customer usability. Conveying "who" the system was and whether she liked tennis or golf was not going to help the caller get what he needed. The focus on the VUI and the persona began to detract both the designers and the organizations deploying self-service speech systems from the goal of maximizing usability and understanding where the role of call automation fit in the overall solution. The result was an unnecessary placing of calls into silos.

Experience has helped us recognize that deploying successful speech applications means striving for a balance of automation and customer satisfaction, which requires a focus on the end-to-end experience. Rather than over-focusing on the persona and the branding—both important— we're recalibrating our efforts to find ways to get a complete, holistic view of the caller's end-to-end experience.

When someone picks up the phone and calls the bank or a catalog company, there are

...a balance of automation and customer satisfaction...requires a focus on the end-to-end experience

many components working together, from the telephony to the customer service agents, contributing to the "customer experience." To continue to build upon our successes with speech recognition applications, we need to look beyond our area of immediate focus, that is, the IVR, the speech engine, etc. We need to put ourselves in the caller's shoes and see past the individual components of the contact center so that we can focus on the "experience" as its own entity.

No technology, no automation, is ever 100% effective. That being the case, it's important that we focus on the entire customer experience— designing not only the automated portion of the interaction, but advising and consulting on the events pre- and post- call routing and automation as well. What seems like "little things" to the companies deploying self-service applications are often "big things" that affect caller satisfaction— and the caller's patience. One of the most common complaints made by callers is the frustrating and irritating need to repeat information. Callers give account information or birth dates in the application only to be transferred by the IVR to a live agent where the same information is asked again. Only viewing the customer experience in terms of a speech interface misses the key factor to success, which is to focus on the entire customer experience and usability.

For those of us working in the area of speech recognition, we need to perform more as "Contact Center Consultants" rather than as VUI Designers or Speech Scientists. While self-service IVR applications are a very large part of most corporate contact centers, we need to assess the caller's needs and goals and make recommendations accordingly. As Contact Center Consultants, we need to help organizations, our own customers, improve their contact centers by focusing on the following services:

- **Identifying and segmenting the caller population:** What are the demographics? How often do customers call? This will allow us to "personalize" service offerings to the appropriate user population. This may include specialized routing and unique automation services, as well as sharing data gathered during the call with an agent when, or if, they are needed to assist.

- **Determining the most common service requests:** *Why are people calling? Can requests be automated? What should NOT be automated?* One of the most common mistakes organizations make is to continue to add functionality to automated systems that should not be there in the first place. Recognizing how users interact with the system and focusing on offering the right tasks in self-service is the key to user adoption.

- **Determining the best methods, approaches, technologies for maximizing usability:** *IVR or web self-service applications? Speech or touchtone? Should we use CTI to facilitate the exchange of information across the contact center?* Many companies focus on finding a technology and then searching for a problem. By viewing the customer experience from end to end and being able to apply the right technology to the right business problem, we can deliver hard returns on investment while maximizing user satisfaction. Similarly, being able to build a solution that seamlessly interoperates allows an organization to intelligently use the information known or gathered about the user to better serve their needs. It also allows organizations to offer proactive services such as outbound solutions and alerts by any media that are preferred by the end user.

- **Establishing an on-going "feedback loop":** *How are the automation rates? What are our callers saying and are they happy? What can we do next to further improve usability?* By incorporating all that we've learned about successfully deploying speech applications into the larger sphere of the corporate contact center, particularly with respect to the importance of a holistic customer experience design strategy, we create a win/win situation. The organizations investing time, money, and effort into improving or creating an effective contact center will see a strong return on their investment. Self-service automation significantly reduces handling costs for companies; and the more calls that can be automated, the fewer customer services agents are required. Most customers today would prefer to handle routine tasks with an IVR than speak with an agent. There is no doubt that there are some tasks that will always require human intervention, but getting tasks done quickly, efficiently, and with confidence is always preferred. Happy and loyal customers are the result of well-designed and well-maintained self service systems.

In summary, VUI design has demonstrated and proven its importance, but the industry is evolving rapidly and for too long has focused on VUI design in isolation. Speech recognition and VUI design are only a part of the overall solution. For organizations to optimize their contact center operations, they have to concentrate on the caller's start-to-finish experience, creating a Contact Center Design, if you will, in order to address the complete customer experience and exploit multimedia technologies available to today's consumer.

A Verbal Exchange

David Ollason

David Ollason, lead program manager, Microsoft Speech Server (MSS), **Microsoft***, discusses some VUI design decisions in a release of Microsoft Exchange, which will have bundled speech recognition. David's career in speech started 16 years ago with British Telecom Research, working on the, then, relatively new area of sub-word unit modeling. He went on to join a start-up, Entropic Cambridge Research Labs, spun out of Cambridge University U.K., and worked on Version 2.0 of HTK (HMM Tool Kit), and later, led the VUI Applications team, building speech interfaces for a variety of applications. The company was acquired by Microsoft in Nov '99, and David continued to lead the Speech Applications team, developing sample applications for MSS 2004 and MS Connect, the auto-attendant currently deployed at Microsoft. He subsequently moved into the role of program manager, responsible for the "authoring experience" in MSS 2007 and VUI design for the Exchange Unified Messaging application, and is now Lead Program Manager for both Authoring and the Platform. He holds several patents in this area, and obtained an MEng in Electrical and Electronic Engineering from Heriot-Watt University, Edinburgh.*

The current release Exchange supports Unified Messaging and provide users with voice access to their calendar, email, personal contacts, directory, and with this release, voicemail too. In addition, there is also a corporate auto-attendant. This application promises to be one of the most widely deployed speech-enabled applications to date, and I'd like to discuss some of the design considerations behind its VUI.

> "Exchange...promises to be one of the most widely deployed speech-enabled applications to date

Complexity is the Enemy – A Simple, Consistent Core VUI Please

Considering the speech-enabled messaging scenario, one is immediately tempted to think big and facilitate interactions like, "Schedule a meeting from 8 to 9 with John Smith." This would indeed be an amazing interface, but would only work outside of the demo scenario if: a) the system were truly capable of recognizing the myriad of possible multi-slot and ambiguous inputs, *and* b) that the caller is fully aware of what can be said. Although today's technology goes a significant way towards facilitating the former requirement, it's the combination with the problem of caller-awareness that makes this kind of interface potentially confusing and unworkable. The example above is ambiguous as to AM/PM, but, more importantly, creates an expectation in the mind of the caller that all possible input forms are supported, thereby eliminating many potentially useful design-time constraints. Instead, we chose to center the VUI design for this application around some simple metaphors, and to branch out to provide more advanced capabilities in a controlled manner.

From a VUI perspective, let me split the application services into two groups; calendar, email and voicemail in one group, and personal contacts and directory in the other. Although the services within each group differ slightly in their operation, each group has a consistent VUI underpinning it: Calendar appointments, emails, and voicemails are all presented to the caller in the form of a navigable list. The caller hears an abbreviated, context-sensitive list of commands for the current item and they can move between items in the list, or issue a command that applies to the current item.

However, the list metaphor is not appropriate for Personal Contacts and Directory. Here the items are usually known to the caller ahead of time and far larger in number. Direct access is therefore a much better metaphor in this case. The caller is asked to speak the name of a contact, which is potentially confirmed and/or disambiguated, and then the caller is presented with a list of the pertinent options for the selected contact.

Simple, but not Simplistic - Compelling Features Required!

Being relatively simple in its core operation does not have to lead to an application that is pedestrian or lacking in features. Quite the opposite— an easily understood and strong foundation is the ideal framework upon which to hang a rich set of features. In addition to the core dialog described above, the application supports being able to recognize items from the Active Directory, to record and send messages as voice attachments to email, and to redirect the call. From this basic framework and set of capabilities a variety of interesting features are provided, throughout the application. The caller can:

- Connect directly to the meeting organizer/room, to the sender of an email/voicemail, or to a named contact in either the Active Directory or Personal Contacts.
- Find emails from colleagues or forward emails or meetings to colleagues, distribution lists, or friends.
- Record voice messages, and attach them to various forms of messages, such as Reply(All)/Forwarded Emails or Meetings,

Cancelled/Declined Meetings, or Voicemails, or attach them to messages to colleagues, distribution lists, or friends.

Other options include:

- I'll Be Late – the caller can specify the number, or a range, of minutes they expect to be late by, and a formatted text email is sent to all the meeting attendees.
- Clear My Calendar – the caller specifies either a time of day or a number of days and their calendar is cleared up to that point.

With a large feature set such as this, one of the key VUI challenges is to effectively educate the naïve caller as to the available options. Additionally, because this is a Business-to-Employee application, and likely to be used frequently by the same individual, the VUI must also cater to the expert-user.

Give Them What They Want, When They Want It

With a serial interface such as a VUI, it is important to present the information to the caller in the most effective order. This is particularly important if a key usage scenario is for the caller to triage the information quickly, as opposed to digesting every detail—callers want to be able to decide as quickly as possible if a particular item should be skipped, cancelled, deleted, read, etc. As such, the email header is organized to present the salient information in the following order:

<If unread> <Sender> <Subject> <Arrival Time>

Similarly, a Calendar appointment is delivered as:

<If tentative, If conflicting, If your meeting> <Time> <Location> <Subject>

With barge-in fully enabled throughout the application, this organization allows the caller to make the earliest possible decision for each item.

At the end of each item presented in a list (of meetings, emails, or voicemails), the VUI plays out an abbreviated, context-sensitive list of the most likely commands. For example, if the current meeting is already in progress then the VUI plays:

SYS: "…entitled Team Meeting. You can say Next, Call Location, I'll Be Late, or More Options"

If the meeting has not yet started, *Call Location* is swapped for *Call Organizer*. Here again, the VUI attempts to maximize information delivery by only providing, by default, those options most likely in the current context. In all cases the abbreviated list ends with *More Options*, through which the caller can hear the full set available.

It is good practice to use careful judgment when deciding whether or not confirmation, with its delay to the flow, is warranted. In many cases the decision is clear cut; we must confirm the name for the *Forward Message* command or risk forwarding someone's message to the wrong person. In other cases, it's more subtle; recognition of *Reply All* is confirmed but *Reply* is not, and in some cases either offering an Undo operation or simply allowing the caller to perform the operation again is actually preferable. For example, in the case of *Delete*, the caller is given a hint, once during any single call, explaining how to undo the operation. The action of *Delete* is to place the email in the Deleted Items folder, and is therefore not as destructive an operation as one may first think.

They'll Talk, Even If You Don't Understand

It's important to note that while the VUI attempts to present only a subset of commands by default, all commands are always available, even if they don't make sense in the current context. It is rarely, if ever, the right idea to remove commands from the grammar in contexts where they seem to be inappropriate. Despite the potentially rare occurrence, the results of an out-of-grammar utterance (poor recognition, or worse, misrecognition) will almost certainly leave the caller in a confused state. Much better is to recognize the command and inform the caller of why it is inappropriate. As examples, the caller can say, *"Cancel the meeting"* even for ones that they did not organize, and, callers requesting to speak to executives via the Auto-attendant, are understood and politely directed to the Operator.

Beware the Can Of Worms...And the Demo Feature

Although the VUI is grounded in simplicity, it does allow multi-slot input in a few key cases, and employs after-the-fact educational hints in order to inform the caller of this functionality. The following kinds of shortcut are supported:

> *"Calendar for tomorrow"*
> *"Find by name – John Smith"*
> *"Forward this to John Smith"*
> *"I'll be ten minutes late"*

For these scenarios, if the caller originally selects the single-slot approach (e.g. *"Calendar"*, followed by supplying the day separately), the dialog gives the caller a hint:

SYS: *"Thanks, and by the way, you can save time in the future by saying 'Calendar for tomorrow' at the main menu."*

We think that this method strikes the appropriate balance between allowing the callers to drive the VUI effectively, whilst not encouraging them to ascribe a greater than warranted intelligence to the system.

Now, you may well ask, "But isn't a phrase like 'cancel my 2 PM' just the same as the ones above?" On the surface it appears to be about the same level of complexity, however, there are important differences here. It would break the fundamental VUI model by allowing actions to be performed on a named item – consider the following worms escaping, "cancel my meeting with John Smith" and "cancel my team meeting," where there may be more than one contact by that name or more than one team meeting. The objects for the supported shortcuts are well defined entities (Day/Date, Name, N-Minutes), whereas "cancel my 2 PM" opens the door to ambiguity.

In Summary

The Exchange Unified Messaging application does not claim to break new ground in terms of VUI design. Rather, it focuses on the consistent use of tried and tested techniques to produce an application that is relatively simple to drive, but also feature-rich, and therefore both useful and usable. This is particularly important, given the wide variety of potential deployment scenarios and differing user groups.

The Amazing Gap

Moshe Yudkowsky, Disaggregate

Moshe Yudkowsky, Ph.D., president, Disaggregate, goes back to basic principles—what we really want is a personal human assistant—to discuss what the objectives of voice-interactive systems should be. Disaggregate, a consulting organization, was founded in 2002 by Dr. Yudkowsky, author of The Pebble and the Avalanche: How Taking Things Apart Creates Revolutions. *He began his work in speech technologies at AT&T Bell Laboratories, where he led a team to design and develop AT&T's automated operator. He left Bell Laboratories in 1996 to join Dialogic, a manufacturer of equipment for the telecommunications industry. At Dialogic, in his role as Senior System Architect for Speech, Dr. Yudkowsky worked with speech technology providers from all across the world. In parallel — at both Bell Labs and at Dialogic — Dr. Yudkowsky was Chair of the Automatic Speech Recognition Task Group of the Enterprise Computer Telephony Forum, an industry standards organization. Dr. Yudkowsky helped found the Midwest Speech Technology Association and is a former board member of AVIOS, the Applied Voice Input Output Society.*

If you call my office and I'm not available, what do you want as an alternative to speaking to me directly? During a recent talk I asked a roomful of people what they would want, and most offered the idea that they'd like to either find me or leave some sort of message.

But that's not what they really wanted—after all, why did they call? They called to find out if I'd received the document they'd sent me; what I thought about it; when I'd get back to them about the document. Since I wasn't there to answer the phone they wanted to know when I'd be available to discuss it.

But what do we offer, as voice user interface designers? Voice mail. Voice mail doesn't even come close to satisfying the actual reason for calling me.

Now let's look to another telephone transaction, a phone call to your bank. Again, why did you call? Maybe you called about your mortgage: you want to know if the paperwork is complete, how long it will take to come to a decision, what the bank thinks of your application, and how much money you're likely to get. What does the bank offer? For the most part, an IVR system that may allow you to check your balance. In fact, some credit-card IVR systems insist that you listen to your current balance and payment information whether you want to hear about it or not.

Finally, an example from everyday, non-telephony life. When you go to the shopping mall and park your car, what do you want? When I my audience asked for suggestions, they responded quite clearly: they wanted to be able to find their car when they exited the store, they wanted their car to actually still be there, and they wanted the radio to still be in the car. What does modern-day technology offer? Flashing lights and a noisemaker, which is a far cry indeed from what the customer wants.

This gap between what customers want and what our applications provide is something of a mystery, and when faced with a mystery, it's best to go to an expert. I refer of course to the Queen of Mysteries, Agatha Christie, who said something quite interesting: "I never expected to be so poor that I couldn't afford a servant, or so rich that I could afford a motor car."

Agatha Christie was born when servants were the norm, and in fact people sought jobs as servants as a far better alternative to life on the farm or in factories. For a bit of amusement, see if you can find an old copy of an Emily Post book on etiquette and read the section that describes a properly-staffed upper middle class residence. An upstairs maid, a downstairs maid, a "tween" maid for the jobs in-between, some kitchen maids, a cook, a butler, footmen, chauffeur, gardeners... an endless list of employees.

Agatha Christie lived through the period when "things"—manufactured goods—became cheap but people became expensive. I lived through this time as well; I recall doctors who made house calls for the very sick, before doctors' time became far too expensive. For that matter, doctors used to work alone in an office with a receptionist and a few nurses; now doctors work in groups so they can share the burden of paying for staff. And remember when automobile mechanics used to fix parts instead of simply replacing them?

With this insight in mind, let's revisit the three different scenarios I mentioned earlier. When you call my office and I'm not there, you connect to my voicemail system. If that system is particularly clever—rather, particularly brilliant—it might be able to read my personal calendar to find out where I am and when I return to my office. And the system might also be able to hunt through my list of contact numbers to find me and transfer the call. But regardless of all that, for the foreseeable future, no automated system can tell you what I think about any particular document or what I think must be done next. To do this, you must either ask me or ask the missing person who used to work for people like me: the personal assistant, the man (who else but men worked in offices one hundred years ago?) whose sole duty it was to manage my day and free me up to work on those tasks that only I can do.

When I call my bank, the obvious person to speak to is a personal banker whose job it is to answer my questions. I'd be delighted if that personal banker was at my beck and call—but today only the most wealthy individuals merit that level of service. The personal banker is gone, and I count myself lucky if the person who answers the phone can tell me if the name of the bank changed that week. Regardless of the efforts of your bank's voice user interface designer it's infinitely easier to turn your head and say, "Please send George a check for $30" to your personal assistant—a five-second utterance—than it is to talk to any IVR system in existence.

Finally, my shopping trip to the mall. I don't want flashing lights and a noisemaker; I want a chauffeur. Actually, I also want a footman or two to carry my packages for me. Speaking of Agatha Christie, let's not forget her contemporary, P. G. Wodehouse, and one of his most famous fictional characters: Jeeves, the gentleman's gentleman, who would no doubt accompany me on my shopping trip to help choose the proper attire. (Today, perhaps only Bill Gates can afford a gentleman's gentleman, someone who carefully picks exactly the wrong color sweater and makes certain that the sweater is somewhat rumpled but not *too* rumpled.)

What are we to do as voice user interface designers to close the huge gap between what customers want and what we offer?

One theory holds that we need to re-create "Mabel," that is, create voice user interfaces that mimic the plugboard operators. While this position certainly has a great deal of merit, I contend that Mabel isn't smart enough. Mabel the plugboard operator might know where everyone in town is and their habits, but that's just plain nosiness (and more than a little creepy in some ways). A modern version of Mabel would have to include deeper knowledge of your schedule. For example, if it's wintertime, a Tuesday or Thursday morning, and there's snow on the ground in Chicago, don't bother calling Moshe—he's outside cross-country skiing. Mabel simply wouldn't know that.

Allow me to offer an alternative suggestion to Mabel. When trying to improve a service, don't look at how it's offered today; instead, consider how that service was offered twenty, fifty, and one hundred years ago.

Voicemail provides an excellent example. Today we use voicemail to take messages. Twenty years ago, large businesses used a call center—a pool of clerks who took messages on behalf of everyone in the company. Call centers were a replacement for the secretary who took messages for a small group of employees and handled some of their affairs. The group secretary, however, was a replacement for a person who was commonplace one hundred years ago: the personal assistant, who provided full time support and attention to your affairs at the highest level.

I therefore contend that if you spend your time improving the voicemail interface, you're improving a cheap replacement (voicemail) for a cheap replacement (call centers) for a cheap replacement (group secretary) for the service that people really want (personal assistant). I won't pretend that current technology can possibly replace the personal assistant—we'd need artificial intelligence, and there's no indication that artificial intelligence will be available any time in the next twenty years. But as long as the voice user interface and application designers consider what people really want—the basic function that we're replacing—we can be clever and perhaps even offer drastic improvements.

If this seems a bit far-fetched, ask a teenager to show you his mobile phone. You'll discover that incoming calls have distinctive rings based on who's calling (as in the personal assistant's "Sir, your friend Mr. James is on the line, would you care to take the call?") as well as a photographs of callers ("Ah, yes ma'am, it's the young gentleman you met at Ms. Deidre's soiree last Thursday night.") and in some telephony systems even the ability to reject calls based on caller's ID ("I'm sorry, sir, but Ms. Susan is not at home.").

These services seem new to us, but actually they are part of a trend: Bits and pieces of the personal assistant continue to come forward to the present day from its era in the distant past, driven by customer demand and

Bits and pieces of the personal assistant continue to come forward to the present day from its era in the distant past, driven by customer demand and providing excellent profits.

providing excellent profits. We can do better, and people want more, than a cheap replacement for a series of cheap replacements.

Talk, Don't Touch

Marcello Typrin

Marcello Typrin discusses why mobile phone users prefer to talk rather than touch. Typrin is Director of Product Management and Planning at Tellme (a Microsoft subsidiary) responsible for the company's mobile and telephony experience. Prior to joining Tellme, Marcello ran Product and Services Marketing for Nuance Communications.

When it comes to mobile phones, people today think about flicking the screen rather than talking into the receiver. The traditional user interface for the phone appears to be rapidly moving away from using your voice to using your fingers. But based on recent Tellme studies, 80% of people would rather talk to their phones than touch them.

Why? It's simple. It takes four touches and 20-odd keystrokes to find an average business with the iPhone, while it takes as little as one push of a button and one verbal command to find the same business with a voice user experience like the one coming from Tellme this fall. Sending a text message with the iPhone takes five touches, one scroll and up to 160 keystrokes, while it can take only two button pushes and two verbal commands to send the same text using our voice experience.

Why is this important? Everyone is waiting for the iPhone "killer." However, we believe it isn't another phone that will settle the score.

> "...new voice experiences across phones will be the user interface that levels the playing field"

Rather, new voice experiences across phones will be the user interface that levels the playing field by solving a common user need – multitasking – that touch interfaces ignore. In fact, recent research conducted by Tellme reveals that an overwhelming majority of respondents said they would feel comfortable using voice to perform a range of tasks on their smartphones while walking (93%), exercising (92%), and shopping or running errands (87%).

The mantra of mobile phones over the past decade has been "do more." First it was email, then Internet. Now, voice experiences are poised to drive the next "do more" wave by making it easier to multitask with your phone.

Consumer demand for a great multitasking experience, we believe, will drive a willingness of the mobile phone industry to incorporate voice. Combine that consumer demand with the mobile operators' business plans and recent advances in voice technology, and voice will be a first class citizen on phones again.

Making Voice Synonymous with Data: Feeding the Wireless Business Model

You may remember the website Kozmo.com, a crowd pleaser that let you order groceries and videos online and have them delivered in minutes. The convenience was great, but, like so many potential ideas, the business model just didn't work.

One of the advantages of voice is that it solves a clear consumer need and supports the mobile operators' business model. As anyone in the wireless industry knows, the growth for mobile operators like Verizon, AT&T, or Sprint isn't in making phone calls (i.e., voice plans). It's in signing up customers for data services. Carriers are building their business on the increased demand for data and the prices they can charge. A $40-a-month call-only plan can exceed $100-a-month when the subscriber signs up for an unlimited data and text messages.

By addressing common (yet highly underserved) multitasking scenarios, mobile operators can expect subscribers to use voice interfaces to send more text messages, increase web browsing, and boost their search activity resulting in data plan upgrades as well as acceleration of advertising-based revenue streams.

The results validate this premise. Over the past several years, Tellme has implemented voice experiences on a range of mobile phones and the response has been overwhelming. We've seen impressive user adoption and repeat usage with approximately three out of every four tasks initiated by users speaking the request, not by entering it via the keypad.

> "...impressive user adoption and repeat usage with approximately three out of every four tasks initiated by users speaking the request, not by entering it via the keypad"

Plus, voice can be a powerful differentiator for operators and services. In our research, 75% of people said they would choose a phone that allows them to get information simply by speaking, rather than typing or using a touch screen. Think of how AT&T has used the iPhone to attract customers to its touch screen. A voice-driven phone could draw the same attention and consumer demand.

Voice Sounds Better

Admittedly, the biggest challenge for us is getting people to try it again.

While voice command applications for mobile phones have been available for years, they have earned a reputation for being inconsistent and hard to use. In the past, the user had to "train" the application to understand what it wanted, improving its accuracy and reliability each time. Users would have to dictate up to three separate versions of an entry so the speech recognition could accurately sample the way it was pronounced and the unique characteristics of the user's voice.

Another challenge has been that the tasks that voice command applications supported were not especially relevant to the average user. Dialing a contact using your voice is certainly important and useful, but issuing a simple voice command to 'launch calculator' is not particularly interesting or relevant to the way we use our phones. What we need is a voice experience that lets us compose text messages, dictate emails, and speak our search queries.

A third challenge was processing power. In order to perform tasks with our voices, an on-device speech recognition solution is not adequate. These are computationally intensive speech recognition requests requiring horsepower that today is only available in the network. The good news is that the rise of phones with fast data connections and in-network speech recognition processing is available to make these experiences a reality.

What all of this means is that speech technology is catching up to the precision of typing. As the technology has become more sophisticated, users can say what they want and get it on the first try. Demands for multi-tasking will lead to them to try it again.

The Strange Road Voice Has Taken

When Tellme was founded in 1999, we had one goal. Bring the Internet to anyone, over any phone.

That has led us in a lot of directions, from the migration of large-scale phone services, from proprietary applications to open standards applications, handling billions of calls annually for enterprises and carriers nationwide, and even our popular 1-800-555-TELL phone service.

Most exciting of all our work may be the return of voice to the phones, but it won't be as we think of voice on phones today. It just so has it that a byproduct of the "smart" developments that made phones powerful multitasking devices will also succeed in making voice technology one of the biggest tools of the phones of tomorrow.

VUI: The Next Generation

Dave Pelland, Catherine Zhu, and Julie Underdahl

Dave Pelland, Catherine Zhu, and Julie Underdahl, Convergys, address the opportunity created by the multimodal capabilities of wireless devices. Dave Pelland is the Director of the Design Collaborative at Convergys. He has been designing, implementing, and usability-testing user interfaces for over 20 years. Since 1993 he has been actively involved in the voice user interface industry. Starting with voice as an original team member at Wildfire, a pioneer in virtual assistant technology, Dave has worked on a wide variety of applications from enterprise solutions to large-scale telecom deployments for carriers such as Orange and France Telecom. Dave has Bachelor and Master of Science degrees in Computer Science from the University of Lowell. Catherine Zhu and Julie Underdahl are both members of the Convergys Design Collaborative led by Dave. They have been extensively involved in the research and design of multimodal applications at Intervoice.

Technology advances, notably 3G GSM mobile networks, have made simultaneous voice and data connectivity available to the masses in a way it has never been before. For those of us designing interfaces for the mobile phone, this multimodal technology has created a new world of endless possibilities.

With multimodal, we're referring to the ability to interact with an application using multiple input mechanisms throughout the application. It's not just adding speech to a graphical interface or vice versa. Rather, it's an approach where a user has the choice of using either modality to complete their task in a single session. The specific type of multimodal applications we've been designing run on mobile phones where users are interacting by using their voice and a graphical user interface on the handset at the same time.

the ability to interact with an application using multiple input mechanisms throughout the application

Much of what we've learned in designing speech applications can be applied to multimodal applications. However, there are additional design considerations introduced with multimodality. What follows is a discussion of some of the new design situations we've encountered and the progress we've made.

Barge-In

Barge-in has become a basic feature of speech interfaces. If someone speaks over a prompt, the application stops talking and acts on the user input. It matches natural conversation well. A multimodal interface adds the new element of interrupting the voice application using the display and vice-versa.

In our applications we've been treating barge-in by either modality the same and have kept the screen moving forward in sync. A response is a response. It gets tricky when you've got multiple slots and multiple inputs on the screen, but keeping the screen in sync with the data gathered by speech is important.

List Navigation

List navigation is one of the first features people point to when extolling the benefits of multimodal interfaces. Having the screen allows for much quicker navigation of lists versus reading through them one by one. The trick is in figuring out what to do with the speech user interface when the list is on the screen. It seems logical to ask the question "Which widget did you want to buy?" But what about retries? Timeouts? If a user is navigating using the screen, you don't want to keep asking for speech input.

While this may seem like a good spot for turn-taking between the different modalities, that doesn't appear to be the case. While most callers will prefer to scroll through a large list and look at them, we're finding that some people still prefer to speak their choice after seeing it in the list.

We're also looking at ways of using multimodal capabilities to share information where prompts might play an abbreviated list but more detailed information would be simultaneously available on the screen. We expect to see the reverse as well—where users would use voice search to find an email but will still prefer to read it on the screen.

One key principle we've established in our work early on is to keep the interaction going in such a way that the user is always aware that both modalities are available.

Error Handling

The types of speech recognition errors are pretty straightforward. You get an error for no input and no match. Typical escalation strategies include asking with more details and letting the user know about keypad equivalents. Now that the screen is available, the question is whether it should be used in those strategies. Another question is how the prompting should react (if at all) to errors that occur in the graphical view? Our approach has been that both modalities must compensate and be used to help the caller recover.

We've been using the different modalities as part of each other's escalation path. For example, a user struggling by voice can be directed to give it another try on the screen. On the display, we can offer more options than are typical with voice alone. We can provide options allowing them to switch completely to the display, switch to DTMF, etc. Conversely, a user struggling with the screen can get asked additional questions to help clarify options.

Prompt and Command Wording

Another task that presents a new way of thinking is prompt writing. We knew how users were interacting with the system so we could use phrases like "when you *hear* the one you want" and "please *say* yes or no". Now we need to use generic phrasing that's natural whether the user is interacting using voice or the screen such as "when you *know* the one you want" or "*here's* the list".

We've seen the importance of having the application make it clear that information and interaction are always available in both modalities. We've done this with the generic phrasing, but also by referring the user back and forth with phrasing such as "as you can *see*…" and "take a *look* at…" Even when the caller is in their car and can't look at the screen, we've made it clear that the information is available visually.

Primary Interfaces or True Multimodal

Some multimodal designs use one interface as a primary means of interaction. Examples of this are voice interfaces that use commands like "show me" or "let me see it" to switch from voice to graphical. At that point you interact with the handset until you hit an 'okay' button. Typically there are still some global commands active by voice at all times, but the dialog flow is really driven by one interface at a time.

We're finding that, because of environment and personal choice, it isn't safe to assume a modality, even if the user seems to be primarily interacting in one way or another. We've seen users switching back and forth as they move in and out of their cars, for example. The speech or graphical portion can't just be a backup plan; either modality needs to be able to handle the task at all times.

Clearly some data types like maps or other graphical information can't realistically be presented by voice, but an effort needs to be made to communicate the information. For example, if someone steps out of 3G range, a good design could still make the step-by-step directions available to the user.

Fallback Conditions

Typically in a speech application, the worst that can happen is dropping the call, terminating the application. And with speech applications, some designs handle noisy environments and drop back to keypad prompts, but even that's pretty straightforward. With multimodal interfaces, a whole new set of conditions are introduced that require good design to handle appropriately. What happens if the user drops out of 3G range and the graphical view is no longer available? What happens if they get in their car and are limited to speech input? Each of these conditions needs to be identified and handled, resulting in a robust application that never drops the user. This really does reaffirm the fact that each modality must be designed to handle the complete task at any point in time.

An Opportunity

Technology has once again presented us with an exciting new opportunity. Along with the constant improvements in speech recognition technology, we're now also working with rapidly improving and expanding network technologies and increasing handset capabilities. We really believe the next generation VUI is going to be part of a multimodal application.

We've learned a lot with each application we've built and we continue to research, prototype, and test new ideas in this space. The new challenges we face as designers are easily offset by the additional opportunities available to us with multimodal applications. These capabilities will allow us to improve and expand existing speech applications and also make new applications possible. Think of it as one more tool we can use to achieve our goal of designing the best possible user experience.

Speech Interfaces for Mobile Phones
Mike Phillips

Mike Phillips, co-founder and CTO, Vlingo, discusses issues in adding a flexible speech interface to mobile phones. Mike has been active in the speech technology world for over twenty years. He started his career as a researcher first at Carnegie Mellon University and then at the Spoken Language Systems group at MIT working on core technology for automatic speech recognition. In 1994, he founded SpeechWorks based on technology that he and others had developed at MIT. In 2003, SpeechWorks was acquired by ScanSoft (now named Nuance). Mike joined ScanSoft as CTO and oversaw technology integration and development across the product groups. In 2005, Mike left ScanSoft to spend a year as a visiting scientist at MIT before starting Vlingo in the summer of 2006.

Mobile phones are increasingly becoming people's personal information, entertainment, and communications devices. As mobile devices, operating systems, and data networks become more advanced, there are an increasing number of applications that are not only possible, but in fact best suited for use on a network-connected mobile phone. As an example, the GPS-based navigation software available on mobile phones is now better in many cases than much more expensive solutions built into high-end cars—because these network based systems can be constantly updated with new map and route information, and will increasingly have dynamic content such as real-time traffic information.

Given these advances in devices and applications, the user interface is increasingly becoming the fundamental limitation on the functionality that can be made available to users. The mobile industry is

> The mobile industry is already grappling with the twin problems of discoverability and usability of applications given the small form factors of mobile phones.

already grappling with the twin problems of discoverability and usability of applications given the small form factors of mobile phones. The prevailing solutions tend to make use of a series of menus (perhaps arranged as a set of tiles on a screen) to allow users to navigate through series of steps, choosing among a small number of choices arranged on the screen. This of course results in either providing a small number of functions, or causing the user to go through many selection steps. For example, finding the weather for a particular location on a Blackberry requires seven steps—and it's well known that these devices have some of the best user interfaces in the industry.

We started Vlingo in the summer of 2006 with a focus of applying speech technology to significantly improve the usability of mobile devices.

Speech Interfaces on Mobile Devices

Many people agree that speech input should be able to help with both the discoverability and usability of applications—it really is the only interface modality that is not subject to the form factor constraints of mobile devices. But, what is the right way to use speech technology to help with this interface challenge? The obvious answer is to take the approaches which have proved to be successful in the IVR industry for providing automated customer care over the phone, and apply them to the applications on mobile phone—perhaps making them "multi-modal" by making use of the display and/or keypad.

We decided instead to start from the needs and constraints of the mobile phone ecosystem and view the challenge of how best to improve the usability of mobile devices, rather than from the speech industry perspective of how to best apply known speech interface approaches to mobile phones. It may be a subtle distinction, but has led us to a very different set of solutions than we had first anticipated.

From looking at the mobile phone industry, we determined a set of needs:

- **Consistency across applications.** Just like personal computers, mobile phones will eventually have large numbers of applications available, and we think all of them should be able to make use of a speech interface—just like they can all make use of the keypad or other buttons on the phone. If the speech interface works differently across these applications—for example, imposing application-specific constraints on what can be said where—users will not be able to efficiently use this wide set of applications and adoption will be limited to only the most popular applications.
- **Don't take away existing input methods.** Even though we know speech is a more efficient way of entering text, there are plenty of situations where users are not going to want to speak (due to privacy, or due to high noise environments where the speech interface may not work well enough). Therefore, we need to augment existing input methods such as keypad and touch-screen input rather than replacing them.

- **Minimize application changes**. Although we know that the user interface is a key success factor for applications and devices, the developers of mobile applications have many other priorities and constraints—including having to run across many different operating systems and devices. So, to achieve market adoption of a new interface modality, we cannot require them to make significant changes to their applications. There are not many providers of mobile applications who are willing to rewrite or re-structure their applications around the constraints of a speech interface. So, having to reorder the flow to ask certain questions before asking others (such as having to ask city/state first in order to use a constrained grammar speech recognizer to obtain a business listing) is just not realistic.

These factors have led us to a very simple notion—provide an enhanced text entry method that works in a consistent way across applications.

Multi-Modal Text Entry

Current mobile phone application environments (J2ME, Brew, and the various smartphone operating systems), already provide application developers with some built-in functionality for collecting user input – so to draw a text-box on the screen and allow the user to enter and edit text using a combination of keypad, navigation buttons, or in some cases a small qwerty keyboard or touch screen.

We have extended these existing text entry methods with speech input. So, the user can continue to use the keypad and other modes of entering and editing text, or at any point, can hold some designated key (either the "Talk" key, or a side button on the phone – depending on device), speak whatever they want, and the words appear at the current cursor location – just as if they had typed the words.

This has the advantage of not forcing users to learn new ways of interacting with applications – the applications themselves can work as before (although in many cases, the easier text entry allows application developers to be more ambitious in their use of unconstrained input – adding a search field rather than forcing users to find things through a series of menus for example). In addition, this means that application developers can now support speech entry with little or no change to their applications.

Unconstrained Speech Recognition

While the advantages of this approach are clear to the users and to the application developers, it of course puts a significant burden on the underlying speech recognition technology. A key aspect of providing this consistent interface is to allow users to speak anything they want into applications and have this converted into text—and to not make them feel that there is a set of constraints on what they can and cannot say.

Fortunately, we are able to make use of certain aspects of mobile phones to solve this challenge. Most importantly, they are personal devices that tend to be used by one or a small number of people. Also, with software on the mobile device, we are able to know which application is currently in the foreground and which text field is currently selected by the user. In some cases, we may be able to obtain addition contextual information such as the GPS location of the phone.

So, while what we present to users is that any user can say whatever they want into any application, we are able to make use of all this contextual information (user, application, field, etc) to better model what particular users may say into a particular application and how they may say it. In addition, by displaying these results back to the user in the multi-modal text field and by allowing the user to correct the text, we provide the user with a way to enter whatever they would like (even in the case of a misrecognition), and can use this in a feedback loop to improve the speech recognition for the next time this is spoken.

At Vlingo, we use a client/server model where the client software on the device handles the user interface, audio recording, etc. and sends the audio along with the contextual information to a set of servers. The servers implement both a very large set of recognition models, and most importantly, a set of adaptation techniques which constantly adapt the recognition models to better recognize user input given the contextual information.

We are finding that this is not only successfully allowing broad input across all applications, but also outperforming more constrained approaches even for constrained applications.

Applications

We and our application partners are using this approach on a wide range of applications—including things such as music or content search, destination entry in navigation systems, and even in truly open tasks such as open web search and messaging applications. By making use of existing application paradigms—users enter text and the application operates on this text—users immediately know how to interact with the applications through the combination of existing text entry methods and speech entry.

More recently, we have found that we can extend these techniques to not only allow filling text boxes with text, but also to provide top-level control of a mobile device (so, solving the discoverability problem). In this case, we allow users to press and hold a designated speech key at any point, say whatever they want, and we use this to route to an appropriate application, filling in the text boxes of that application as if the user had used existing input methods. We think that this combination of unconstrained text entry into applications combined with this overall flow between applications will become a very compelling way for users to interact with applications in a wide variety of mobile devices.

Will it Blend? The Multimodality of Speech and Text

Eric Collins

Eric Collins, Vice President of Mobile Marketing, Nuance Mobile, discusses the interaction of user interface modes. As Vice President of Mobile Marketing for Nuance Mobile, Collins is responsible for all activities related to mobile strategy, product marketing, deal support, market development, communications and M&A. Previously, Eric held the role of General Manager of the Tegic Business Unit, where he was responsible for the full product set from Tegic including T9 predictive text, XT9, and T9 Nav. Eric received his JD from the Harvard Law School and his bachelor's degree from Princeton University with a major in Public and International Affairs and a minor in African American Studies.

It seems as though the mobile device market has been evolving and growing at breakneck speed over the last few years—particularly with the advent of the iPhone. We're relying on our phones more than ever to not only make calls, but to send text messages, emails, download music, get directions to a coffee shop we also just found by searching the Web—it's amazing, and it's only going to continue to get wilder with the downloadable application market rapidly heating up.

But just because we have an amazing pool of applications and services at our disposal to get our devices to do just about anything, getting access to those apps faster and easier is a key challenge for handset manufacturers and carriers.

… we have an amazing pool of applications and services at our disposal to get our devices to do just about anything, [but] getting access to those apps faster and easier is a key challenge

Outside of the clear need to ensure the user experience is the best one for their consumers, there is a huge revenue potential at stake. Easier access to and increased usage of applications and services in many cases ultimately increases the average revenue per user (ARPU). And it's all about ARPU, right?

Predictive text and speech recognition software are already playing a role in creating that path to the ultimate user experience, and many of the phones coming to market today have both predictive text for faster and easier texting, and basic speech capabilities like Voice Activated Dialing. But this caliber of implementation is just scratching the surface of how powerful a multimodal input approach can be for creating an amazing user experience that drives revenue and brand loyalty. And of course, there are a number of external factors, like distracted driver legislation, that clearly demonstrate the value of input options being fully integrated.

The Keypad Evolution Kick-starts Multimodality

The evolution of the phone keypad is a good example of how we're beginning to see increased multimodality on mobile phones—whereas most traditional consumer phones used to only include a 12-key phone pad, the full QWERTY keyboard made its way beyond the BlackBerry and into more and more feature phones. Largely driven by texting and the evolution of lower-level devices being able to now handle email, consumers demand faster and easier text input capabilities. With that in mind, we've seen many devices over the last few years that feature a 12-key phone pad on the main interface for easy dialing, and a slide-out QWERTY for easy texting. Consumers are able to get the best of both worlds—easy calls, easy texts.

But with advent of the iPhone came the touchscreen user interface (and of course, the App Store phenomenon)—and a new era in mobile phones was born. It's amazing to think that the Apple iPhone was introduced in January 2008, and a year and half later, handset manufacturers including Samsung, Palm, and RIM are already following in line with their own touchscreen devices that feature both full QWERTY and 12-key inputs. In just the last two or three years, we've seen an evolution of modality on the keypad. And there is a clear market for both—with mobile users addicted to the look and feel of the touchscreen, and those who wouldn't dream of giving up their hard QWERTY keypad.

Moving Beyond the VAD

As mentioned earlier, many of today's phones have basic voice recognition capabilities for dialing numbers, finding contacts, and checking the status of your phone's battery, eliminating the need to scroll through address books and menus. Voice input is perhaps the fastest and easiest input method, which is why there is an increasing adoption of expanded voice command and control capabilities. In fact, Datamonitor recently released a report stating that advanced speech recognition technologies in mobile devices will triple over the next five years, rising from $32.7 million in 2009 to $99.6 million in 2014. We're seeing this traction unfold, with many of today's higher-end feature phones now enabling users to access just about any application or service on the device with their voice. Looking to play BubbleSmile? Just say "Go to

BubbleSmile" and you're there. Need to send a text message to John Smith? Just say "Send text to John Smith," and the input field appears.

This is where the multimodality of speech and text input are increasingly intersecting, and where the myth of the multimodal handsets is starting to become a reality. For instance, Samsung devices like the Instinct and the Memoir will allow users to pull up the text input screen with their voice, and automatically bring those users into a touchscreen QWERTY text input field that features predictive text. And of course, there are devices coming to market with full dictation capabilities where you launch the text application with voice, dictate the message and then send it with a simple voice command—no text input at all. This completely voice-enabled phone is not far at all from hitting the market, and the technology is here today, where consumers can surf the Web, download music and get directions—all via voice commands.

So why the need for a complementary input solution when voice-driven access is the fastest and the easiest? Simply put, there are many situations where speech input isn't appropriate. If you're in a meeting, and you need to send a message to your boss that you're about to close a major deal, traditional texting is clearly more appropriate. Or if you're at a concert and want to download the new song your favorite band just played, you'll likely have a hard time with the voice query in an incredibly noisy environment.

Similarly, there are many situations where text or manual input just doesn't work either—namely while driving. Many governments across the globe are passing legislation that mandates hands-free control of their phones as a means of thwarting their ability to text while driving, and keep their eyes on the road. And for good reason. A 2008 study conducted by the Technical University of Braunschweig in Germany found that utilizing speech to even dial a number on their phone improved the ability to maintain the ideal car position by 19% when compared to manual dialing. Better still; speech input was also approximately 40% faster in making a call, reducing the distraction period by the same amount. With the fully speech-enabled handset, drivers have the ability to interact with their device as needed.

Multimodal Output

Today's mobile consumers are quickly realizing that multimodality is more than just input—it's also about output and the ability to manage services. Voicemail to text services are perhaps one of the best examples of this, where voice messages are intercepted by an automated service or team of transcriptionists to transcribe a message that is then delivered straight to the user's inbox or sent as a text message. Gone are the days of having to listen to a voicemail, then write down a number or message, or worse, missing a voicemail if you can't access your phone—it's all completed in a matter of minutes for you and dropped into your inbox.

With regards to output, text-to-speech technology is increasingly being integrated into today's devices to confirm an interaction with the phone succeeded or failed, read back text messages, confirm phone numbers and contact information, read aloud directions, and more. Text-to-speech integrated with speech input is the key to the truly hands-free device, where consumers are able to have a conversation with their phone to do virtually anything, such as download a song or get directions. This is one of the most exciting advancements in multimodality that we're going to see more of in the coming months.

It's also worth noting that similar output exists for the hands-on experience as well, with hard and touchscreen keyboards providing haptic feedback, a vibration or sound to confirm that the interaction had taken place.

Speech and text are increasingly becoming intertwined and integrated on the phone as part of the best possible user experience that gives consumers the ability to access what they need when they need it, which in turn is driving increased brand loyalty and revenue opportunities for OEMs and operators.

A key aspect of multimodality, however, that cannot be lost is keeping that 'power to choose' input in the users' hands. Each user's experience is their own, so it's important to not completely dictate their interaction with the device and what input method they need—some will likely rely heavily on speech as more phones come to market with advanced capabilities. And of course, there are those who often find themselves in situations where text input is a must-have. It's all about what the user is trying to achieve, and ensuring they have the utmost innovative input technology to get there.

Future-Forward Multimodality

Multimodal input and output is an exciting area to watch, as additional areas of speech technology become closely integrated with mobile applications and services. For instance, imagine combining interactive voice response (IVR) technology with enhanced output on the device. One example is calling 1-800-FLOWERS, where the IVR options are enhanced on the screen of the device, providing pictures of the flower options with pricing information, or the ability to design your own flower arrangement right from your phone. To help ensure consumers continue using the service, 1-800-FLOWERS would remember what you ordered in the past, your preferences, etc., and enable the user to place similar orders easier and faster, using voice or text input, right from the phone.

This scenario is a culmination of on- and off-device applications intertwined with a variety of input and output options that not only drive value for OEMs and operators, but also for the companies that sign on for them.

There's no doubt that multimodality will continuously evolve as it provides a richer user experience, enabling consumers to leverage the input and output methods that are best suited for them moment to moment.

Developing Unimodal and Multimodal User Interfaces

James A. Larson

Jim Larson, Speech Applications Consultant, Larson Technical Services, discusses how to coordinate a GUI and VUI design that might be used as alternatives or together. Dr. Larson is author of the home study guide and reference, "The VXMLGuide," www.vxmlguide.com. He is co-chair of the W3C Voice Browser Working Group, program chair for Speech Technology Conference East in New York City, and columnist for Speech Technology magazine.

Developing user interfaces for a unimodal user interface, such as a graphical user interface (GUI) or a voice user interface (VUI), is a difficult task. Developing a multimodal user interface is even more complex. This chapter presents a strategy for developing a GUI and VUI for the same application that, while different modalities, are similar in terminology, structure, and behavior. These user interfaces can be combined into a multimodal user interface in which users select the mode that is most appropriate for the task at hand within the current environment and subject to the user's preference and background. The approach consists of four major steps:

a strategy for developing a GUI and VUI for the same application that...can be combined into a multimodal user interface in which users select the mode that is most appropriate

1. Develop a user's conceptual model for the user interface.
2. Develop a GUI and a VUI that conforms with the user's model.
3. Refine the GUI and VUI to make them easier to use.
4. Integrate the GUI and VUI into a multimodal user interface.

To illustrate each step of our approach, we will use an example of an online catalog sales application where the user browses the catalog and selects items for purchase by placing them into a shopping cart.

Step 1. Develop a user's model for the user interface

After writing down several possible usage scenarios, the designer creates a user's conceptual model. This captures the essence of the user interface without getting bogged down with details like screen layout, interaction objects, and other creative aspects of user interface design. Instead, it captures the major application objects which the user will manipulate, the attributes and relationships of these objects, and the operations which users are expected to perform on the objects. Most importantly, it captures the *flow control*—the possible sequence of operations that users may perform.

Figure 1 illustrates a typical flow control for a user interface for shopping consisting of a catalog that users browse, and a shopping cart where users place goods and services to purchase. Users apply operations (represented by arcs) to the catalog, which includes searching for a specific catalog item, viewing the next catalog item, viewing the previous item, and placing items into the shopping cart. Users review items to be purchased by moving from the catalog to the shopping cart, and resume browsing the catalog by moving from the shopping cart to the catalog. The users check out at any time. (To keep the example simple, we do not include the checkout process.)

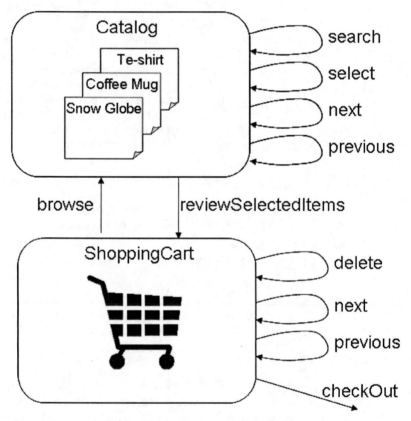

Figure 1. Flow Control for the Catalog Sales Application

Step 2. Develop a GUI and VUI that conforms with the user's model

Figure 2 illustrates a prototype graphical user interface with buttons representing the operations shown as arcs in Figure 1. By clicking the "Next" and "Previous" buttons in Figure 2, the user views items in the catalog. By pressing the "Shopping cart button," the user opens the shopping cart window.

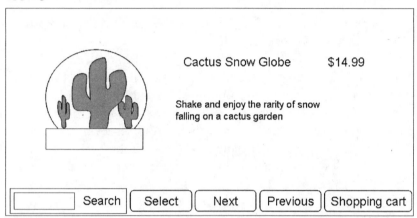

Figure 2. GUI for the Catalog Sales Application

Figure 3 illustrates a prototype voice user interface with voice menus representing the operations shown as arcs in Figure 1. By speaking menu options, users browse the items in the catalog. By speaking the option "shopping cart," users move to the shopping cart portion of the dialog.

Computer: Shake and enjoy the rarity of snow
             falling on a cactus garden, only $14.99

Computer: Do you want to search, view next, view previous,
             or go to the shopping cart?

User: Shopping cart

   ...

Figure 3. VUI for the Catalog Sales Application

Step 3. Refine the GUI and VUI to make them easier to use

Revise the GUI layout, colors, fonts, and graphics to make the user feel comfortable and able to use the GUI easily. Revise the prompt wording, grammars, and sequence of fields in the VUI to put the user at ease when using the VUI.

Step 4. Integrate the GUI and VUI into a multimodal user interface

Because both the GUI and VUI use the same user's conceptual model, the same business code, and the same terms and operations; it is possible to integrate the GUI and VUI into the multimodal user interface. The user can read and hear the same information and may click/type as well as speak to enter information. By using the multimodal user interface for a few minutes, users will naturally choose between clicking/typing or speaking for each operation. Users choose different methods of entering data different times depending upon environmental concerns (noise pollution, low light, small keyboard). If for any reason, one user interface becomes difficult to use (e.g., low light for reading the screen or high background noise that users must speak over), users can select the best mode for the situation.

For example, it is probably faster for users to read the contents of the shopping cart than listen to the options via voice. However, it may be easier for users to speak the names of desired items rather than to visually browse through a big catalog. For efficiency purposes, it may be desirable to emphasize either the GUI or VUI over the other for specific tasks. However, keep both interfaces available for every task in the event that users need an alternative interface due to external environmental constraints or simply because users prefer one interface over the other.

At some point during the design process, the flow control for the GUI and VUI may begin to deviate from each other. For example, the GUI may provide a sequence of items which users can purchase, while the VUI may simply ask the name of the item that users want to purchase. When this occurs, the GUI and VUI interfaces diverge, so more work is required by developers to present consistent information to users of each user interface. In this case, users also have greater difficulty applying what they know about one user interface when they attempt to use the other user interface. For simple user interfaces, it is possible to manage this divergence. For complex user interfaces, divergence could become problematic. Multimodal user interface designers should strive to strike a balance between maintaining a consistent flow control for both the GUI and VUI, yet tailoring each user interface to take advantage of its unique capabilities.

Why the GUI and VUI should be similar

There are several advantages in keeping the GUI and VUI as similar as possible. Similar GUI and VUI user interfaces enable users to:

- Choose either the GUI or VUI depending upon which device is available. If users are in an office with a desktop computer, then they should use the GUI. However, if users are in a car with only a cell phone, then they can use the VUI. Users should be able to switch between user interfaces even while using the application, such as leaving the office and walking to a meeting in a different part of the building.
- Select the user interface with which users are most familiar. If the device supports both GUI and VUI, then users can select the user interface they prefer. Many users will select the graphical user interface because they feel comfortable with a mouse and keyboard. Others will select the voice user interface because it leaves their hands free for tasks.
- Select the user interface most appropriate for the current environment. If in a business meeting, users will select the GUI to avoid noise pollution during the meeting. If the environment is noisy and the speech recognition system is likely to fail, users also may select the GUI. If the lighting is bad or their eyes are busy, users may select the VUI.
- Switch between the user interfaces at will. Users perform exactly the same operations in both the GUI and VUI. Both the GUI and VUI use the same commands and work consistently, even though the user interfaces are very different. Users apply their knowledge of how one user interface works when they use the other user interface.

This flexibility is possible because both user interfaces access the same database and the same underlying application-specific code. Only the types of user interfaces are different.

Multimodal User Interface standards

The World Wide Web Consortium (W3C) Multimodal Interaction Working Group has developed an architecture to support multimodal user interfaces [1]. A principle component of this architecture is the Interaction Manager, which is responsible for coordinating modality components such as GUIs written in XHTML and VUIs written in

VoiceXML 3.0. A candidate language for specifying the Interaction Manager is State Chart XML (SCXML) [2], being developed by the W3C Voice Browser Working Group. Using the W3C multimodal architecture, developers could use SCXML to specify both a data model and flow control, XHTML to specify the GUI, and VoiceXML to specify the VUI.

[1] http://www.w3.org/TR/mmi-arch/
[2] http://www.w3.org/TR/scxml/

Multimodal UI Guidelines for Mobile Devices

Ali Mischke

Ali Mischke, User Experience Manager, Vlingo Corporation, discusses the integration of speech technology with other user interface modes available on mobile phones. Since joining Vlingo in November, 2007, Ali has been responsible for user interface design, usability testing and user research. Prior to her time at Vlingo, Ali was Senior Manager, Analysis and Design, Iron Mountain Digital. At Iron Mountain, she founded the user experience discipline and was responsible for all user research, user interface design, and usability testing across the company's Digital product lines. Ali has 12 years of experience as a user experience designer on a broad range of products and domains—mobile speech recognition as well as digital cameras and digital projectors, web-based content management, email archiving, PC backup and dialog marketing. Ali began her career as a Usability Engineer at Eastman Kodak and holds a Bachelor's degree in Linguistic Anthropology from Brown University.

Against the backdrop of smartphones and mobile application stores, the expectations we place on mobile devices are multiplying exponentially. SMS and web search are givens, and are quickly being joined by the need to email, get directions, read restaurant reviews, locate friends, buy movie tickets and complete a nearly unlimited array of other activities from the essential to the frivolous. As both the power and the complexity of mobile functionality explode, we become confounded by a consequent lack of discoverability and usability. With so much at our fingertips, how can we identify what functions are available to us, access those features and finally figure out how to use them?

Traditional solutions involve display-based interactions or pure voice-controlled interfaces. Unfortunately, both models are flawed. Display-based interactions constrain users to the number of menu items that can appear on-screen at a time, which is simply no longer enough. On most handsets, navigation and text entry can be tedious and requires more focus than users are often willing or able to dedicate. On the other hand, speech-only interactions are inefficient for complex tasks like messaging, and there are many situations when speech is either not desirable or not possible.

Considering the advantages of each mode of interaction, the right solution must be multimodal, in which users can speak or type into a text box and receive both audio and visual feedback when speech recognition occurs. Speech is both faster and easier than typing, providing a natural solution to the problems of discoverability and usability. Additionally, because multimodal text boxes offer a choice of input method, the approach adapts comfortably to suit users' changing contexts.

The case for convergence

The matrix of decisions involved in multimodal interface design is dizzying, yet imagine if users needed to learn a new set of rules to type in each mobile application. The task would quickly overwhelm users into inactivity. Convergence of multimodal UI design is equally critical to unlocking the power of mobile devices.

Multimodal user interface guidelines can be considered at two levels: that of the text box and that of the application as a whole. At the application level,

> Convergence of multimodal UI design is critical to unlocking the power of mobile devices

there are some domain-specific best practices, but each application should and will implement its own workflow. However, convergence at the text box level is both essential for mobile usability and achievable without overly constraining application design. In this column, we present a selection of mobile interaction guidelines for multimodal text boxes.

Usability at Vlingo

At Vlingo, user experience design is an obsession fueled by field usage data as well as findings from well over 1000 usability and beta testers. In June of 2008, Vlingo launched the first voice-controlled idle screen client for BlackBerry, allowing users to search the web, voice dial, send messages and more. The product has since been released for iPhone and Symbian devices, and was in beta for Windows Mobile at the time of this writing. In a survey of 80 Vlingo BlackBerry users, Vlingo's ease of use was rated 4.4 out of a possible 5 points. In the spirit of continuous improvement, each major product release includes extensive design iteration driven by user research and usability testing.

Multimodal text box design

Multimodal UI design at the text box level must address the following questions:

When is speech available?

The key to multimodal interaction is flexibility—allowing speech or typing at all times so the user can choose, situation by situation, the most appropriate input method. If some text boxes within an application are speech-enabled and others are not, users must learn and remember where they can and cannot speak. This model invites user error and diverts attention from the task at hand. Furthermore, users who are initially hesitant about speech recognition are more comfortable if they know that typing is still available. For maximum simplicity and power, multimodal text boxes must allow the user to speak or type at any time.

How do users know they can speak?

Hold and speak Effective speech cues must be clear, concise and context-appropriate. To this end, the Vlingo TALK hint appears in an empty speech-enabled text box. The text provides clear instructions on activating voice, while the handset-specific icon shows which key to use. The consistent appearance of the TALK hint reassures users that the text box will behave like any other Vlingo-enabled text boxes they have encountered previously. Finally, the cursor reinforces that typing is still available.

How do users activate speech?

The preferred key is device dependent (side key, TALK key, dedicated voice key or virtual button in the case of touch devices); so, too is the interaction model. With regard to interaction model, there are three options: press and hold during speech; tap at the beginning and again at the end of speech; and tap at the beginning, with automatic endpointing. Automatic endpointing is interesting for its simplicity from the end user's perspective; however, due to issues around handling background noise, we believe that it should be used only in conjunction with another model.

On BlackBerry and feature phones, press and hold is the clear preference. In a 20-user A/B study, press and hold proved superior by all measures: user preference, number of errors, severity of errors and user ratings of perceived speed and accuracy. For iPhone, the answer is less clear, perhaps due to the emerging nature of the platform; however, tap at the beginning and end of speech appears most consistent with users' mental model and with iPhone UI standards.

How do we provide feedback?

A multimodal approach allows us to provide a rich set of visual and auditory cues. When users press the voice key, Vlingo provides two types of feedback: an ascending tone to imply "start," and a "Listening..." dialog that appears on screen. Likewise, we mark the end of recording with a descending tone and a transition of the dialog to a "Thinking…" display.

Additionally, we provide a results tone when recognition results are displayed, or an error tone as appropriate. Any network-based application will have some latency; in an eyes-free situation, or when the user may have moved attention elsewhere, the results and error tones guide the user's attention back to the application.

Finally, in situations where the application may be taking automatic action, such as auto-dialing, text-to-speech provides important feedback about what the system is about to do.

How do we display results?

IVR-descended mobile applications show how the system interpreted speech but do not show exactly what the system heard. In case of misrecognition, the user's only choice is to speak again.

Consider a user requesting "sushi" in a local search application. The system might have heard "shoes" and would return the names of local shoe stores. A user seeing the name of local shoe stores when searching for sushi is unlikely to make the phonetic connection, and may not understand why the system is displaying local shoe stores. Without the ability to type or select from a list of alternates, the user loses confidence. Why trust that the system will hear correctly if it made a mistake the first time? After a second or third misrecognition, the user may abandon the system.

In contrast, Vlingo shows recognized text in a standard text box along with the system's interpretation of that text. The recognized text clarifies how the system chose its match, making any misrecognition appear less random. The presence of the text box allows users to correct in the mode they prefer—speaking again, choosing from a list of options, or using the familiar keypad interaction. This correction ability, particularly when paired with an adaptive loop that enables Vlingo to learn from successes and errors, increases user confidence in a voice-based system.

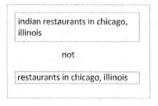

Text box height poses an additional challenge. Imagine that a user says, "Indian restaurants in Chicago, Illinois." In a one-line text box, "Indian" will have scrolled out of view. This is acceptable when a user types, as users know what they have entered.

In the case of speech, however, we must consider the potential for misrecognition. In the one-line text box, users cannot differentiate between the case where "Indian" has scrolled out of view and the case where the recognizer clipped the audio and did not capture the word "Indian." In most cases, users assume recognizer error. To increase confidence and simplify the review process, we allow a multimodal text box to expand dynamically to accommodate all recognized speech.

How can users correct misrecognitions?

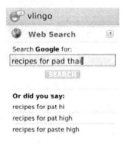

For short utterances such as web searches, users prefer to speak again or choose from a list of phrase alternates. Phrase alternates, or recognition alternates for the entire utterance, provide a one-click means of correcting errors. Interestingly, users rate an application that displays reasonable phrase alternates as both more accurate and more intelligent than an application that simply displays recognized text.

When the user's utterance is longer, as in messaging applications, or when only one word in an utterance is misrecognized, providing alternates at the word level offers the most efficient path to correction. Our implementation varies with the capabilities and constraints of each mobile platform; on BlackBerry, word alternates automatically appear in a drop-down when a user scrolls to a word, much like the predictive text drop-downs that appear when the user types.

Providing word alternates in context allows users to see whether the correct word is available, easily select a word alternate, or immediately start typing if the desired word is not present. Tension exists between ensuring that word lists appear only for words that need correction and ensuring that the word lists appear in a timely manner so users notice the lists before they begin typing.

To prevent flashing of word lists while the user scrolls, we introduce a delay before alternates appear. On trackball devices, a delay of 500 milliseconds when scrolling horizontally and 800 milliseconds when scrolling vertically appears to balance these considerations.

With both words and phrases, we provide alternates but do not require the user to explicitly confirm the utterance or dismiss the alternates.

Given the high rate of recognition accuracy, we assume success but make correction easy and available if it is necessary.

Conclusion

As we increasingly ask our mobile devices to take over or expand upon our computers' capabilities, we become frustrated by the limited nature of small screens and keyboards. When implemented in an intuitive and consistent manner, multimodal interaction can transcend these limitations of discoverability and usability. Through convergence in standards for multimodal text box behavior, the mobile industry has a unique opportunity to maximize both power and simplicity in user interaction, delivering on the promise of a mobile user experience that truly serves and delights.

Advertising and the telephone

William Meisel

Bill is the editor of Speech Strategy News and the editor of this book.

The Web changed advertising, as attested in part to the decline of print newspapers and magazines and the growth of Google. Mobile phones are a new arena that advertisers are interested in pursuing, and much of the attention has been on mobile smartphones' web browsers, trying to treat the mobile phones as small PCs. There are at least two problems with this approach: (1) the phone screen is too small to allow the ad-positioned-at-the-side-of-the-web-page model popular on the PC; and (2) it doesn't address the large fraction of users, particularly internationally, that don't have a smartphone (or, if they do, don't want to pay for an unlimited data plan). Since all phones have a voice channel and a microphone, opt-in calls to get information, play games, or contact a company with a conventional call should be harnessed for advertising purposes. Automation is necessary for this approach to be economically feasible because of the possible volumes and because putting a voluntary call on hold is likely to lose the call. Speech recognition can make this automation flexible, even enjoyable, an approach I've called "conversational marketing."

This point has been partially addressed in the chapter on "Trends driving the adoption of speech in mobile and telephone applications" by this author. That section raised the question of whether contact centers would become voice sites, and described some current applications that could be considered conversational marketing, in particular, a phone number one could call during a game show to participate and win a prize, supported by advertising.

Free business directory assistance is one of the more conventional approaches to advertising over the phone. It is modeled on the "Yellow Pages" style of business directory, with companies paying for position when a caller asks for "pizza" for example.

One key to the success of the major advertising model on the Web is that advertisers only pay for a click, and are thus assured that they are paying for an interested prospect. A TV broadcast or ad published in a newspaper or magazine doesn't have this feature. A caller asking for a service or company is the type of qualified prospect for which advertisers yearn. Providing a measurable result with a call is similar to the pay-per-click Web approach.

A call prompted by an ad is of a different nature than a customer service call. The call may be more exploratory, and the caller more willing to spend time on the call. The design of a Voice User Interface for marketing applications may require an adjustment in the usual assumptions about a caller to a contact center.

A call prompted by an ad is of a different nature than a customer service call

If call centers handle such calls, they not only need to be prepared for larger volumes, but for a change in image. Rather than a cost center and a necessary evil, call centers will be viewed as a part of the company's image and marketing effort. Call center equipment and application suppliers will find themselves dealing with a different set of call-center objectives—and find a significant opportunity if they can help their customers fulfill their new role.

Mobile Speech Applications for the University Campus

Bill Bodin and David Jaramillo

Bill Bodin, Chief Architect, IBM Mobility and a member of IBM's Senior Technical Staff, and David Jaramillo, Multimodal Speech Development Technical Lead for IBM Software Group, discuss a multimodal project developed at a university which rapidly became part of campus life. Bodin focuses on innovation at the highest level. Bill has developed solutions and intellectual property in numerous disciplines and is a recognized IBM Master Inventor. Bill was the chief scientist for this university project creating both the architectures and business relationships which were delivered through this endeavor. Bill has been an IBM software developer and innovator for 18 years. Jaramillo has been working on the development and deployment of IBM Multimodal Speech solutions across many different industry segments including automotive, healthcare, education, and field force automation. David has been in software development at IBM for over 15 years and served as a key technical mentor for the project highlighted in this column.

In the summer of 2005, a project began to create innovative speech-enabled mobile software solutions. These solutions, and associated infrastructures, would further evolve to enhance university campus life. The project began as an IBM "Extreme Blue" activity. Within the context of a typical Extreme Blue project, IBM mentors and Extreme Blue Staffers recruit top talent from universities to define and develop innovative use of new technologies. The projects are typically comprised of four students. One or more technical and business mentors from IBM complement the student team. Schedules are very aggressive and the program is extremely intense. Students are unlikely to have any prior experiences with one another and the mentors will meet their student teams for the first time on Day One of the project. The challenge is to take a group of individuals and form a team that will innovate, develop, and deliver a solution in a span of less than 12 weeks. The final project is then presented to the top executives at IBM's corporate headquarters. (For more information on IBM's Extreme Blue program visit: http://www-913.ibm.com/employment/us/extremeblue/.)

CampuSphere

This Extreme Blue project, code named "CampuSphere," was sponsored by Wake Forest University. The student project team was composed of three technical students and one MBA candidate. A team from the Wake Forest CIO staff comprising both technical and administrative disciplines, and a team of architects, developers, and business development personnel from IBM complimented this core student team. The challenge was to define a set of applications that could be designed and implemented in a very short span of time and then be deployed as functioning pilot programs on the Wake Forest campus during the Fall of 2005. The team identified many innovative concepts, but had to quickly narrow them down to a few applications that could be prototyped, developed, and deployed into a production environment. Two concepts were selected as project deliverables: LaundryView and Shuttle Tracking. The team frequently conferred with project stakeholders from IBM and Wake Forest, ensuring project relevancy while making continuous and significant progress.

LaundryView

LaundryView, although simple in overall concept, had its share of complexities. It received much criticism on initial review, but then demonstrated the power of such a system in the environment for which it was targeted. In a university setting, for those living on campus, laundry can be a time-consuming chore. Students often struggle carrying loads of clothing to distant laundry areas only to find all of the machines are already in use. The result is frustration and loss of time that could have been committed to much more productive and fulfilling activities.

The solution was to deploy Internet-enabled washers and dryers from Mac-Gray Corporation (http://www.macgray.com). The solution further comprised the delivery of a speech-enabled system compatible with handheld PDA's and Smart Phones. The networked washers and dryers utilized in this project are capable of sharing status data over the network. These appliances required no further modification in order to achieve network communications as the appropriate sensors, actuators, and networking infrastructure was already provided by Mac-Gray. The systems and methods developed in this project took that network access to the next level by providing a Voice User Interface, or VUI, compatible with handheld devices. This server-based system processed the information gathered from the on-board sensors, transformed this data into appropriate mobile compatible data formats, and delivered this data as a speech-enabled application. The mobile data format utilized in this application was Mobile X+V, or XHTML + Voice. [See, for example, www.voicexml.org/specs/multimodal/x+v/12/.]

This data format combines spoken interaction with standard web content by integrating W3C standards for the visual and voice Web. The two data formats used in this embodiment are XHTML for rendering visual content, and VoiceXML for the spoken interaction component. The XHTML and VoiceXML

Combining spoken interaction with standard web content by integrating W3C standards for the visual and voice Web

modalities are further integrated using XML events. In this way, voice handlers are attached to specific DOM (Document Object Model) events. This application, with text, graphics and voice integration, enabled students to view and query the state of these networked washers and dryers. The VUI, with voice and conventional touch access, combined for an extremely easy-to-use service. Students can specify the dormitory location, and gain immediate access to the status of washers and dryers at that location. The architecture allows for follow-on activities that may include alerts delivered in many possible forms such as SMS (Short Message Service), IMS (IP Multimedia Systems), or Voice alerts. The infrastructure and application was highly successful and was selected for inclusion into Wake Forest's Mobile Application Suite called MobileU. It was also incorporated into the production Mac-Gray LaundryView portal, becoming one of the first multimodal speech-enabled Websites deployed outside of IBM.

Shuttle Tracking

Shuttle Tracking is a system used to locate and track buses that transport students across campus. This required a solution comprised of GPS metrics gathering, transmission of metrics to a central server, and predictive software algorithms which could accurately report a vehicle's arrival time at a specific location. During the early evaluation phase, and in anticipation of the start of the Extreme Blue activities, many devices were tested. These devices included different protocol implementations. While all devices implemented GPS (Global Positioning System) for satellite tracking and location determination, they differed in the methods with which they communicated this information over the network. Some devices communicated using SMS, while others used methods that had less network latency. The team arrived at a decision to utilize a device from Webtech Wireless (http://www.webtechwireless.com) that incorporated global positioning capabilities combined with GPRS (general packet radio services) functionality for transmitting coordinate data wirelessly over a wide area at periodic intervals. The choice of this protocol eliminates the message queuing that is typical in SMS implementations, providing more real-time data access. The coordinate data gathered from these shuttle-installed devices is communicated to a WebSphere application server, which uses location-prediction algorithms and adds the multimodal X+V markup. This multimodal content, compatible with Smart Phones and PDA's, allowed the important voice features to be used on handheld devices. This server-based application was delivered as both a graphics and text-based system combined with a VUI. Students could specify their current location to learn the arrival time for a specific shuttle or query from an extensive list of alternate locations. With WiFi available campus wide, students can simply access this application using their constant connection to the network. They simply specify the location of their choice and gain instant access to the next shuttle's arrival time.

While both projects presented specific technical challenges to the team, the most significant aspect was schedule pressure. A brief, ten-week development runway, however, gave way to two significant and deployable speech applications. These applications are now used daily on the Wake Forest campus, empowering students with mobile solutions which enhance their personal productivity.

In conclusion, this fast-paced project was extremely exciting and considered one of the hottest, most talked about projects in the Extreme Blue program. It was clear to all involved that the voice enablement added an element that made for a compelling easy-to-use application. This voice enablement clearly added elements of accessibility and convenience not normally available in a hand-held platform. In the end, it will be interesting to see the impact of these significant projects as they become increasingly popular with students both on and off campus.

Enabling Personal Device Functionality in the Car

Thomas Schalk

Tom Schalk, vice president, speech technology, ATX Group, discusses some of the appropriate ways speech technology can help ease the use of devices in vehicles. Dr. Schalk leads a group that focuses on developing speech-enabled telematics services. Prior to joining ATX, Dr. Schalk was the CTO of Philips Speech Processing and the CTO of Voice Control Systems. He has over twenty years of experience in the speech recognition industry. He received his Ph.D. from the Johns Hopkins School of Medicine and his B.S. in Electrical Engineering from the George Washington University.

People are learning that speech interfaces fit nicely into the driving experience, particularly when interacting with personal devices. Today, speech interfaces allow drivers to use speech to control music, phones, navigation systems, and other functionality, making it possible to have fun and be more productive while driving. The ultimate goal is to provide speech interfaces that are comparable to human-to-human interaction: Talk to your car the way you want, but focus on driving. Although we are a far cry from reaching the ultimate goal of human-like speech interaction, significant speech advances have occurred over the past few years. In this chapter, we will examine how speech has enabled personal device functionality inside the vehicle. In this context, we will explore what it takes to deliver effective multi-modal interfaces under driving conditions.

what it takes to deliver effective multi-modal interfaces under driving conditions

Bringing speech to the task of driving

Driving is so basic to modern life that drivers don't think of it as a complex task. In fact, driving requires a vast amount of physical coordination as well as analytical skills. Surprisingly, the cognitive load of driving has not decreased over time. Increasing traffic levels, complex mixes of road systems (often subject to construction or constriction), and a much higher flow of information and infotainment to the vehicle make ordinary driving a very demanding challenge. Physically, every part of the body is involved in driving. Even today's most advanced vehicles still require hands on the wheel and feet ready for the accelerator and brake pedal. Even with technological advances, driving is still largely a silent activity when it comes to tools and controls. Speech is not always an easy interface to use, especially in a hands-free automotive environment when others in the car are talking. If a car has speech, it is usually an optional interface mode.

It has only been recently that speech in the car has gained user acceptance, and a leap in public awareness was first spawned by Honda's aggressiveness with the introduction of its 2005 Acura speech-enabled features. And, with the recent marketing activities associated with the Ford-Microsoft Sync product, more people are aware of what you can do with speech and how it makes sense in the car. As evidenced by Fiat, Ford, with Hyundai to follow, it is very obvious that Microsoft and its partners are focused on providing the means to improve the lives of drivers and passengers by enabling personal device functionality and connectivity inside the vehicle. It is anticipated that speech will become a critical component of automotive multi-modal interfaces.

Bluetooth mobile phones in the vehicle

Although speech interfaces are still most prominent in high-end vehicles, the Bluetooth mobile device is changing that. Just a few years ago Bluetooth phones were capable of being paired with cars to get hands-free voice dialing. Once the vehicle detects the phone and synchronizes, the phone audio is channeled through high quality car speakers and the device's microphone is replaced by the vehicle's embedded hands-free microphone. Essentially, the vehicle takes control of outbound and inbound calls and also manages the audio. When the user interface is designed correctly, all you need in the vehicle is one additional "speech" button.

Let's review a few usage scenarios. The driver starts the car. Seamlessly, the phone connects to the vehicle. While driving, an incoming call is detected. The radio mutes and starts playing ring tones. The driver simply pushes the speech button and begins a hands-free conversation with the destination party. To end the call, the driver pushes the speech button. The driver, while listening to the radio, decides to call a friend. After pushing the speech button, the radio mutes and a short greeting is played. The driver says "call Bob" and the call is initiated. The driver also has the option of saying something like "dial 726 555 1234" instead of "call Bob." Depending on the user interface design, confirmation may or may not be part of the dialing dialogue (e.g., only confirm when low recognition confidence scores occur). The driver can also push the speech button and notice that his car display is showing his outlook address book that is accessible from his phone. Knobs associated with the display (e.g., a navigation system) can be used to select the desired party, and the driver can simply say "dial" to initiate the call. Very simple, very slick, and very safe as long as conversations do not cause driver distraction, a safety concern not expected to go away anytime soon.

So, with the introduction of Bluetooth, a new set of drivers were given the opportunity to dial by voice under driving conditions while improving safety. The difficult part for the user remains in teaching your car to detect your Bluetooth-enabled mobile phone. Once past the initial setup and a small amount of experimenting, voice dialing is something people can't live without. It is interesting that voice dialing user interfaces are pretty much the same as they were twenty years ago, but with much better speech technology.

Other personal device functionality: music, navigation, and text messaging

In addition to voice dialing, we are now seeing MP3 song management through multi-modal interfaces that include speech as primary modality. iPods and other media devices are connected and controlled by the car and the device audio is channeled through high quality car speakers. Speech input is handled by the vehicle's embedded hands-free microphone. Say the name of a song, see the name displayed, push a knob to confirm, and enjoy the music. So again, the personal device user interface is changed, once connected inside the vehicle.

It makes no sense for a driver to type in a vehicle, especially with a knob that you twist and nudge until you highlight each target letter, followed by a push (knobbing). Even though it's a very awkward experience, there are cases for which knobbing is the only way to enter a destination into vehicle navigation system, and the vehicle must be stationary.

Navigation is definitely an application in need of a robust speech interface. The ultimate user experience is to say your destination as though you're telling a cab driver where you want to go. Today, there is limited speech input capability (quite impressive at that), but not to where you'll be recognized when you say "take me to an economy hotel with air conditioning within 12 miles of the airport." But at least destination entry is speech-enabled, and it will only get better. We are already experiencing single utterance address input, as opposed to saying individual destination components, such as *city*, *street name*, and *street number* in a directed dialogue fashion.

The portable navigation device (PND) has grown tremendously in popularity and many believe that before long, expensive on-board navigation systems won't be needed in cars as smart phones and other handheld devices with navigation capability become commonplace. Earlier this year, the first speech-enabled PND was introduced by TomTom. Based on personal experience, the ease of destination entry improved dramatically. It is logical to expect PNDs to somehow connect to the vehicle much like phones and iPods do today. The PND experience will include the benefits of the car's audio speakers for voice-delivered turn-by-turn directions, and also the hands-free microphone for speech input. The vehicle video screen (if available) will appear as a large scaled version of the PND display. Combine the in-vehicle knobs, buttons and touch screen features and you have matched a high-end on-board navigation system that is expensive to purchase and maintain with current map and POI data. Take the vehicle-synchronized PND to the voice search level for local search and you've exceeded the best of what you can buy today from a car dealer. One has to believe this will be a reality soon.

Text messaging has rapidly become a mainstream communication method among consumers. And now, *most* mobile phone users use text messaging. Not limited to just the younger generation, even adults admit that they text while driving, which leads to a very dangerous situation unless the user interface is radically changed to accommodate the task of driving. Even today, text messages can be read inside the vehicle via text-to-speech. What is coming is speech-enabled outbound texting made possible by advanced speech recognition. It is easy to imagine an extended voice dialing user interface that handles "text Bob" or "call Bob." How close is the speech technology?

Web browsing in the vehicle

As dangerous as it may sound, the internet will become part of the driving experience. From a high level, the website user interface must be designed with human factors in mind, and with the goal of providing relevant information while minimizing driver distraction. Recent advances in voice search technology suggest that voice browsing will eventually be practical while driving, mixing spoken search requests with visually displayed web content, tailored to the driving environment. Multi-modal interfaces should be aimed at ease of use and should adapt during different driving conditions. It will be interesting to observe how mobile device advertising is brought into the driving experience. Ad-supported directory assistance, web ads, and other channels may need to be transformed to be proper for the driver. Speech is again critical.

In summary, many personal device functions can be enabled in the car with proper multi-modal user interfaces. We have touched on voice dialing, music control, navigation, texting, and web browsing, and social networking is yet another candidate. For each of the examples given, speech stands out as the key enabler.

Speech-Enabled Memory Assistants

Patti Price

Patti Price, Owner, PPRICE Speech and Language Technology Consulting, looks at the features and tradeoffs in mobile services that have the potential to extend our memories and become our personal assistants—at least in part. Dr. Price has over 20 years experience in developing and transferring speech and language technology, including the French telephone company research center CNET, BBN, and SRI International. She is a co-founder of Nuance and two other technology companies, BravoBrava! and Soliloquy Learning. Her formal education includes postdoctoral training in electrical engineering at MIT, a PhD in linguistics from the University of Pennsylvania, and a degree in French literature from the University of Poitiers. She consults in speech and language technology planning, especially educational applications, proposal writing, and patent litigation strategy.

On the lookout for speech applications beyond dictation and 'your call is important,' I foresee growth in applications aimed at consumers and small businesses. Since I have often wanted to upgrade my own memory as easily as that of my devices, I looked at a few speech-enabled memory assistants. I figure that small (and large) businesses, like individuals, also need memory assistance if they want "the right hand to know what the left hand is doing."

Interface designers for such applications face the usual dilemma of providing powerful features while keeping the interface simple. One could separate interfaces into the 'Don Norman' vs. 'Doug Engelbart' approaches. In this simplification, the 'Norman approach' is characterized as "make it obvious" and the Engelbart approach as "make it powerful". In reality Norman has argued that the goal is not simplicity, but rather making complexity understandable. Analogously, Engelbart would argue that no interface should be *needlessly* complex. Both would agree that power is good and understandable is good. In the real world, differences in approaches will show up in the inevitable compromises needed.

Speech interfaces are particularly problematic in that, as individuals and as a species, we have vast experience and expectations about how spoken language interactions work. To quote Don Norman: "Develop a system that recognizes words of speech and people assume that the system has full language understanding, which is not at all the same thing."

236

While these expectations are real, it is also true that people have vast experience accommodating our speech to those who differ from us. For example in talking with children, foreigners, the hearing impaired, etc., we learn to accommodate to differences in linguistic, cognitive, social, and other norms. Therefore, humans can likely adapt to a new, machine-based system and might be willing to adapt more if the potential reward is larger.

Humans can likely adapt to a new, machine-based system and might be willing to adapt more if the potential reward is larger

Several services now transcribe voice mail, with varying degrees of success (e.g., YouMail, CallWave, SpinVox, PhoneTag from SimulScribe, GotVoice, Google Voice). While such services assist memory in making voice messages as searchable as text, I'm looking for more. Many of these seem to be a hybrid: automatic transcriptions reviewed by humans. This is a clever approach that bootstraps current technology while providing data for potential next-generation technology. The use of keywords can trigger a reminder later and is a clever tradeoff of speed for accuracy. Although the pioneering "I want Sandy" closed operations December 2008, I found a few speech-enabled memory assistants currently available.

me2me.com does not transcribe the message itself, but recognizes separate voice tags used as categories.

JustKnow.com does not do any recognition, but leverages text-to-speech so that web information (directions, phone numbers, etc.) can be sent now or later as text or voice to devices or calendars. The service is sold to the website, not the consumer. While this could help jog memory, it depends on information push from the company rather than information pull from the person who needs it.

Jott.com offers several information pull products. *Voicemail* ($9.95/month for 40 messages) transcribes speech via software with human review. The resulting text can be shared with others, searched and organized like regular email. Memory assistance can come through callback reminders or by sharing messages. Jott's *Assistant* (under $4/month or pay per minute) focuses on memory, allowing you to get voice notes transcribed, integrate them with Outlook tasks and other applications, send text messages by voice, set up reminders and calendar events that integrate with other calendars, connect with about 30 web applications by voice (e.g., Twitter, Facebook, Amazon), and get RSS feeds read to you. This definitely should assist personal and corporate memory and information access! Particularly interesting for businesses is *Jott for Salesforce* ($25/month including Assistant). This plan allows you, in addition, to add voice updates (with confirmation) to Salesforce accounts and opportunities, add tasks and notes, schedule appointments and set reminders. The interface is simple: "Who do you want to Jott?" You must know, of course, that the possibilities include 'notes,' 'reminder,' etc. Depending on the answer to the initial question, you may be prompted for the message, or for time and date and then the message.

ReQall.com is intriguing: Don Norman is "Chief Mentor" and Sunil Vemuri, the technical co-founder, did his PhD work at the MIT Media Lab on "Memory Prostheses" (<http://web.media.mit.edu/~vemuri/wwit/wwit-overview.html>). The free version enables adding items by voice or text, unlimited voice transcriptions, organization of items based on keywords (dates, times, buy, note, meet), categorization by time, things, people, and sharing of reminders with others. Free version users get a free month of 'Pro,' which (for $25/year) adds access to the keywords 'at home,' 'at work,' setting up places as keywords, reminders based on your calendar, organization based on categories you define, and location awareness (if your device has GPS). Imagine you're near a hardware store you put on your list of places—you suddenly get a message with your list of hardware store items.

Regarding the interface, Norman's website (<http://www.jnd.org/>) says, "We worked hard to make it really simple, to eliminate all the features that came to mind. No features, therefore no fuss. Simple and powerful." This recalls the Onion News Network spoof of Apple allegedly releasing a laptop without a keyboard—"just one big button -what could be simpler?" Despite this comment, as a fan of Norman's work and Vemuri's idea of memory prostheses, I tried it out. The interface at the top level is a step more structured than Jott's, asking if you'd like to 'add,' 'reQall, or 'share.' The website gives helpful examples, but it is hard at first to remember all the keywords and when things must be in a certain order. Although humans don't require that of you, once you know the rules, it feels like a reasonable accommodation. Keywords include: dates (put item on calendar), 'buy' (create list), and 'remind/tell/ask' (share with others). In my trials, the transcription appeared about a half hour later (Jott says theirs takes 10-20 minutes).

I added one contact ("Gene") and recorded a few items. His name was transcribed correctly once, but more often it was not: twice 'Jean', and once each 'Jim' and 'Keane'. Although proper names are tough for humans or machines, with 'remind' a key word and 'Gene' the only contact, 'Jean' could have been automatically changed to 'Gene', and a request for verification might have been generated for 'Jim' and 'Keane'. The recognition was otherwise reasonable with some annoying exceptions, e.g., "can" misrecognized as "can't." (These are so frequently confused by humans that Noam Chomsky, the linguist, often says "can yes" for "can"). Verification, though tricky with a delay of 30 minutes or more, probably needs to be addressed. I also observed some errors hard to imagine coming from either humans or a speech system with a reasonable trigram language model, e.g., "I need a ground cloth to sheets, a blankets, …"

Summary

Jott and ReQall and others will likely find users. Jott seems poised to address a largely untapped market in organizational tools and information access for small businesses. Their integration with other applications and with Salesforce in particular, should position them well for this. ReQall, on the other hand, aimed at consumers, could particularly appeal to aging baby boomers as memories fade. Both companies, as well as the intriguing Wolfram Alpha (a "computational knowledge engine," <http://blog.wolfram.com>), combine automated with human effort. This is a great idea, but observed error rates can vary with the transcriber you happen to get, which complicates assessment. However, Google's text-based search engine adds value and has been successful despite the notorious difficulty of determining search accuracy; a pragmatic solution should be possible here, too. I am hopeful that these companies can help us achieve Engelbart's vision of using technology to bootstrap our collective intelligence.

Using multiple speech-to-text engines simultaneously for speedier editing

Jonathan Kahn

Jonathan Kahn, CEO, Custom Speech USA, discusses software integrated with Nuance's Dragon, IBM, Microsoft Vista, and SAPI 5.x speech engines. Dr. Kahn is a practicing radiologist in Indiana and company founder. The company works with multiple engines, languages, and vocabularies, and develops enhancements for transcription, dictating, and translating in a variety of fields. The company is a certified partner with Microsoft and Nuance (Dragon), works with a variety of integrators, including ProNexus (telephony) and Wizzard (IBM and AT&T applications). Co-branding, integration and development tools, and custom programming are available. Recent installations include Columbia University Medical Center Biomedical Informatics, NY, and Montserrat Day Clinic, Queensland, Australia.

Custom Speech offers software tools that make it easier to develop applications using speech engines on PCs, applications that leverage the typical speech-to-text or command execution in speech recognition software from Nuance (Dragon NaturallySpeaking), IBM, Microsoft (speech recognition in Windows Vista), as well as other SAPI 5.x speech recognition, such as our Custom Speech speaker-specific SweetSpeech.

Software tools that make it easier to develop applications using speech engines on PCs

This "do-it-yourself" speech engine and toolkit enables a transcription service to create unlimited speech user profiles from day-to-day dictation. As a vendor of Windows-based software supporting workflow management for speech recognition and text-to-speech and of utilities supporting desktop dictation, transcription, and audio conversion, we've seen interesting ways to improve the basic features of the speech recognition software. Three examples of adding value to speech-to-text applications using our SpeechMax software are discussed in this chapter:

- A speech transcription aid comparing speech datasets processed by one or more speech recognition engines, manual transcribers, or both.
- Addressing privacy concerns when human editors review dictation or when transcription is sent for web- or server-based automatic transcription.
- Filling in forms, including interacting with local or web-based databases.

Comparing recognizer outputs for faster editing

With multi-window text comparison of synchronized speech recognition engine output, a human speech recognition editor can identify differences in text output and quickly correct identified misrecognition errors using audio playback. Experience indicates that locating differences in speech engine output can quickly help identify a high percentage of recognition errors, though occasionally different speech engines make the same mistake. Experience also indicates that the more accurate the engines, the fewer the differences, and that about an 85% accuracy rate is required for substantial editing time savings.

For example, if three speech recognition engines are each 97% accurate and have transcribed a 30-minute dictation audio file, a human speech editor can identify many misrecognitions without having to listen to the entire audio. Instead, the editor can use the tab key to advance to the next difference, listen to the audio, and select the appropriate text from a dropdown of engine output. If correct text is not listed, the editor can manually transcribe it. The edited text may be returned to the original speaker for review and optionally further edited. If all speech engines agree, suggesting a high level of accuracy, editing time approaches zero. The application can apply the same synchronized text comparison techniques to documents created by human transcribers (or a transcriber and speech engine). In some cases, a speech engine may recognize unusual words (e.g., technical terms) more accurately than a human transcriber.

The same tools have been adapted to support synchronization of source text and one or more Unicode translations. Using the tab key, the user can navigate sequentially from phrase to phrase, viewing, for example, the source English, as well as highlighted and synchronized Spanish and French manual or machine translation. Using text comparison, it is also possible to compare a synchronized manual and machine translation from a document originally dictated with speech recognition. With audio-linked text in the document window, a supervising translation editor may listen to the original English to determine translation accuracy.

In a related approach, color-coding can indicate not only the presence of differences, but also the number. If dictation audio is transcribed by three engines, clear (no highlighting) indicates agreement by 3 of 3 texts and low risk of error. Pink highlighting indicates agreement by 2 of 3 texts and moderate risk of error. Red highlighting indicates no agreement among the 3 texts and high risk of error. Similar techniques may be employed for synchronized human output or combined human and computer pattern recognition for both speech recognition and translation. It is also possible to compare more than 3 texts.

Privacy and dictation

Privacy of dictated material is often a concern. SpeechMax also supports "ScrambledSpeech." With this feature, a dictating speaker using real-time speech recognition can divide the segments into two or more groups, scramble (reorder) the segments within each group, and send each scrambled mini session file to a different human editor for correction. This division and scrambling limits any one editor's knowledge of the whole content and helps protect confidentiality and privacy. The corrected sub-sessions are merged and reordered into a final, sequential session file for review by the dictating author or a senior reviewing editor. Alternatively, a dictating speaker can have audio segmented, divide and scramble segments, and send the audio to more than one human transcriptionist or speech recognition server for transcription. The application also supports exporting the "divided and scrambled" audio to different transcriptionists for processing in Word or other standard word processor.

Additionally, the "SpeechCensor" feature in Custom Speech software supports speech and text redaction for removal of confidential material. For example, with user text selection, the program replaces patient identifying information, such as name or social security number, by [DELETED] for the audio-linked text and a "beep" for the audio. This "censor" can remove display at the graphical interface level and the XML file level for removal of text and audio metadata.

Filling in forms and interacting with database or Web-based software

Most reports are structured in some way, and forms are a particularly formal way of structuring data. Custom Speech software can be used to integrate speech technology into a form-filling application. The form may be a Microsoft Word template, a database application such as an Electronic Medical Record (EMR) or Personal Health Record (PHR), or a web-based application.

For example, SpeechMax has been used to correct Word templates dictated with Dragon speech recognition. After migrating the recognition results into SpeechMax, the end user or transcription assistant can select (highlight) text, playback audio with or without a transcriptionist foot pedal, and edit text. The SpeechMax editor can migrate corrected text back into the Word form using an Add-In for Microsoft Office created specifically for this purpose by Custom Speech.

Currently, the company is integrating with a web-based medical Continuity of Care Record (CCR). This will support data migration to and from web-based EMR and PHR databases to Word and SpeechMax. Updates may be completed using speech recognition or manual transcription. The modified data can be optionally migrated into a new CCR XML file for upload into a web-based EMR or PHR or saved locally on a PC.

SpeechMax also allows creating a "Talking Form". Using these features, a transcriptionist or clerk can create a structured dictation form in a matter of minutes. This may include audio prompts with a human or text-to-speech voice. The medical, legal, law enforcement, or other user can record audio for each blank. The audio can later be transcribed manually or with server-based speech recognition. The user may also input data with the keyboard, real-time speech recognition, or a bar code reader.

Using "AV Notebook" features, the user can also add graphics and images to the form or other session file, as well as hyperlinks or commands to open a media player or other program. This speech-oriented multimedia session can be displayed on the Full SpeechMax version, or locked and distributed for viewing in a free Reader version.

It's not just the engine

Many converts to using speech-to-text on the PC wonder why more people and businesses don't use it. Part of the answer may be that a complete application requires more than the basic speech-to-text conversion. This chapter indicates some ways in which the basic technology can be converted into a more complete solution.

The Evolution of Speech Technologies in Warehouse Voice Picking

Doug Brown

Doug Brown, vice president of product management & marketing at Datria discusses the breadth of speech technologies used in the past dozen years to voice-enable warehouse applications. The in-depth chapter presents an interesting evolution of the technologies used in a difficult problem that many don't realize has been one of speech recognition's earliest successes, both for vendors and the companies that buy it.

All three generations of technology offer a viable approach to today's supply chain automation, and Doug examines the reasons behind the diversity in approaches, pros/cons and recent trends, basing his chapter on discussions with colleagues as well as his own extensive experience. Doug became involved with speech recognition technologies at the Conversant Systems startup venture at AT&T in the mid-80s. His current work at Datria focuses on an ever-increasing set of packaged speech applications automating mobile employee processes (including warehouse workers). Datria was formed in 1997 as a spin-out of Lockheed Martin, and has been delivering multimodal data collection and field service solutions over the past 10 years. Significant customers include Johnson Controls, Bell Canada, Coca-Cola Enterprises, Energy South, TELUS, and Cardinal Health. Datria has partnerships with SAP, Cisco (e.g., SSN, May 2008, p. 27), and Nuance (among others).

Over the last 10 years, the most widely known speech recognition application for many has been self-service in the contact center. Lately this has been changing, as speech recognition (voice user interfaces) has started becoming more prevalent in consumers' lives. People now talk to their navigation systems, to 411 information services, to their PCs, to the music systems in their cars, their cell phones (voice dialing and text-message creation), to corporate systems for password reset, for Google and Yahoo voice search, and so on. The unique hands-free, eyes-up attribute of a speech interface is becoming increasingly appreciated, especially as people are "always-connected" while mobile and away from a desktop interface.

Voice picking in the warehouse – once a niche market filled with specialty point solutions and proprietary technologies – is another example of companies using speech technology outside of the contact center. While not as well known as some speech applications, it is a burgeoning market with greater than $400M end-user spending in 2008

(*"The Guide to Voice Solutions in Warehouse Environments,"* Daniel Hong, Datamonitor, February 2009).

An unusual characteristic of the voice picking market is the range of varying speech technologies at play – embedded capabilities versus network speech, speaker-dependent versus independent, adaptive technologies versus fixed voice templates, etc. This chapter looks at why different speech recognition approaches exist, and their relative pros and cons. It is important for the reader to know three things:

- A 2009 voice picking deployment can be successful using any of the approaches discussed below.
- With today's technologies, most misrecognition issues are rooted in user behaviors and are not caused by technological shortfalls.
- Regardless of vendor marketing, no speech recognition technology works 100% of the time for 100% of the users.

Voice Picking

Voice picking is the generalized name for speech recognition applications that automate order fulfillment and other supply chain processes occurring in a warehouse. This catch-all includes a range of process automation addressing goods receipt and put-away, order selection (picking), replenishment, inventory management (cycle counting), storage location moves, cross-docking, returns, value-added services (such as engraving your name on an iPod before it ships), yard management, inspection, loading/packing, and shipping/delivery. Voice picking is easily complemented with other data collection technologies, such as scanning and RFID.

A voice picking transaction is relatively simple when compared to other speech-enabled enterprise mobility applications. Workers log into the system and begin receiving instructions on where to go in the warehouse to pick items for order fulfillment. When they get to the right location they speak a "check digit" from a signage, to confirm that they are indeed in the right place. The system then tells them how many items to pick, with the worker confirming quantities as they pick items. Voice picking systems are flexible and deal with deviations such as stock shorts and damaged goods. From a speech recognition perspective, there is a very small vocabulary (grammar) that is typically 60-200 words.

Companies are increasingly investing in voice picking for very tangible cost reductions, improved staff productivity, increased safety, faster employee ramp-up, regulatory compliance, reduced staff churn, and enhanced operations agility.

Companies are increasingly investing in voice picking for very tangible cost reductions, improved staff productivity, increased safety, faster employee ramp-up, regulatory compliance, reduced staff churn, and enhanced operations agility. Accurate supply chains also please end customers who value the right items arriving on time without damage and with the proper paperwork.

Vendors of voice picking solutions
(See chapter for terminology in columns)

Vendor	Speech Recognizer	Speaker Dependent (SD)	Speaker-Independent (SI)	Adaptive
Cadre Technologies	embedded		commercial	manual
CTG	embedded		commercial	manual
Data Systems Int'l	embedded		commercial	manual
Datria	network-based		commercial	automatic
Genesta (SyVox)	embedded		commercial	manual
Itworks	embedded		commercial	manual
KBS Industrieelektronik	embedded		commercial	manual
Lucas Systems	embedded	proprietary		automatic
Naurtech	embedded		commercial	manual
SAE Systems	embedded		commercial	manual
topsystem Systemhaus	embedded	proprietary	commercial	
Vangard Voice Sys.	embedded		commercial	manual
Vocollect	embedded	proprietary		
Voice Insight	embedded		commercial	manual
Voxware	embedded	proprietary		
Wavelink	embedded		commercial	manual
Total--		4	13	

Pioneers (the 1990s)

The earliest challenge for successful use of speech recognition in warehouses was the level of noise – both steady state (conveyor systems, production lines) and dynamic spikes (forklift honks, pallets being abruptly dropped onto concrete, etc.). In the early 1990s, commercial speech recognition technologies were not robust enough to provide accurate performance in such noisy environments.

Pioneering vendors addressed this issue by simplifying the speech recognition effort. By taking a fat-client speaker-dependent (SD) approach, spoken input (utterances) would only need to be matched against one-person's established voice templates. This simplified the computing requirements and narrowed the possible outcomes. The small footprint of the recognition software and compact computing requirements were an excellent fit to the more rudimentary mobile devices available in the 1990s.

Given the lack of commercial market alternatives in the 1990s, voice-picking vendors were forced to become speech R&D operations, creating proprietary speech engines. While market options would change over the years and commercial alternatives would emerge, these proprietary speech engines can still be found in today's markets, and are still valid alternatives for customers to consider.

In the 1990s, the strength of using SD technology was that quality speech recognition became possible in noisy warehouse environments, creating the voice picking market. SD algorithms were also an excellent fit to limited CPU and memory resources of the proprietary rugged mobile devices of that era, units that preceded the commercial handhelds available today.

Yet SD technology was not a panacea. Recognition against a user's voice templates was very specific, which is to say narrow. If users spoke differently as they tired later in their shifts or had colds, they would not sound like their original recorded voice templates and recognition problems could exist. One vendor, MCL Technologies, created a tool to identify SD users having recognition problems (MCL-Voice Manager). This tool helps solution administrators identify workers having recognition performance difficulties, and allows taking corrective actions. Misrecognition issues were often behavioral (such as yelling instead of speaking in the same voice in which the templates were recorded), but when problems were technology-based, the user had to re-record their voice templates. A recent customer case study published about Carib

Sales noted that all their workers had to re-record their voice templates. The original voice samples were said to be "robotic" whereas actual field use had a more relaxed voice sound. This highlights the need in the SD approach for workers to speak in a manner consistent with their stored voice templates.

According to industry pundit Judith Markowitz (J. Markowitz, Consultants) Lucas Systems became the first of the pioneering SD vendors to use an adaptive speech model approach to avoid re-recording of voice templates. Markowitz notes that, "An adaptive approach modifies the user's voice template over time, adding new acoustic model information through actual use in the warehouse." Adaptive speech recognition, as discussed below, is an excellent technique for improving speech recognition performance in an automated, unsupervised manner.

Larger customers perceive one aspect of SD technology to have hidden costs: the creation of voice templates for each worker. The requirement to have a worker spend 30-40 minutes to record their voice template initially sounds benign. Yet customers with larger user populations and multiple sites have become sensitive to the management resources involved, the lost hours of productivity, and the lengthening of rollouts that delay reaping cost reductions.

For example, Mike Jacks, Senior Manager of Logistics and Transportation Systems at Coca-Cola Enterprises (CCE), recently noted the difference in using SD and SI technology approaches in their multi-location roll-out to more than 2,300 pickers at 100 sites. If SD technology were used, CCE would have had a warehouse supervisor or IT manager involved with each worker's creation of voice templates. Assuming 40 minutes for each picker and 40 minutes for a supervisor, the total investment would have equated to 383 days (more than a year) of lost staff time. By using SI technology instead, the solution worked out of the box for all workers and no time was lost to creating voice templates.

More importantly, deploying voice picking to 100 sites with SD would have taken far longer than the 16 months it took with SI technology. This is a critical point as a longer deployment means delays in CCE realizing the compelling cost savings. In comparison, a recent press release on Morrison's, a sizable UK grocery chain making a self-described "ruthless deployment" of SD voice picking (requiring each user to create voice templates) says that it will take two years to deploy only 30 sites. This is one reason why larger customers have become sensitive to the economic impact of choosing an SD technological approach to voice picking.

One other limitation of SD technologies – and again, smaller customers may be as less sensitive to this – is that SD is only a good fit to warehousing applications, The limitation to small vocabularies (up to 200 words due to the requirement for each user to record voice templates) means SD may not be suitable for cost-savings applications elsewhere in the company. Optimizing other enterprise processes typically requires the ability to recognize vocabularies (grammars) with hundreds or thousands of words: a requirement best met with speaker-independent (SI) technology. Examples of optimizing processes with voice applications beyond the warehouse include: plant maintenance, enterprise asset management, field service management, transportation management, sales force automation, regulatory audit/inspection/reporting, crisis management, and human capital management.

Examples of early pioneering vendors taking the speaker-dependent (SD) approach are Lucas Systems, Topsystem Systemhaus, Vocollect, and Voxware (via the 1999 acquisition of Verbex Voice Systems). All still offer and deploy SD solutions today, with Topsystem becoming the first vendor to offer a choice of SD and speaker-independent (SI) technology in 2007. In the past year, Topsystem (also known as Top-VOX) said at the recent ProMat '09 event that almost all of its new customers have chosen their SI engine, but that it's nice to have the SD technology on hand in case there is a rare speaker better suited to that technology.

Another pioneering vendor, SyVox (originally Speech Systems Incorporated) took a different tack, attempting to move acoustic inputs from a device client application to a centralized server, where speaker-independent (SI) processing would be used to understand spoken inputs. This was a highly innovative approach – possibly years ahead of its time – and parallels can be seen with today's 3rd-generation VoIP approaches discussed below. SyVox also became one of the first vendors to switch from proprietary to commercial speech engines in 1998. Ultimately, the SyVox brand was acquired by Genesta where it is sold today as a second-generation, commercial speaker-independent (SI) solution.

Exploiters – 2nd Generation (2002-2007)

Investment in the commercial speech recognition market increased, especially in the area of small footprint embedded automatic speech recognition (ASR) engines. New applications drove significant R&D and new mobility applications – catapulting speech recognition as the interface-of-choice for speaking to portable navigation systems, talking to your car's music system, dictating text messages, obtaining 411 information, dialing by name and most recently, doing multimodal Google and Yahoo searches. While unrelated to the voice picking market, many of these mobility applications faced similar challenges – how to accurately interpret spoken input in a noisy environment.

This resulted in a range of commercial speech recognition vendors – Nuance, IBM, Microsoft, Loquendo, SVOX and others – producing commercial off-the-shelf (COTS) products. These new engines made it far easier for application companies to introduce voice recognition solutions. Academia also made speech engines available via open source licensing. The impact of new speech engines was quickly apparent on the voice picking market, where a new software-only business model appeared and the number of competitors quadrupled.

In the 1990s, due to immature speech products, voice picking vendors had to invest R&D into proprietary engines (discussed in "Pioneers," above). Once commercially viable engines became available, it was possible for software application-only vendors to enter the market. And many did, notably Cadre Technologies, CTG, Data Systems International, Itworks, KBS Industrieelektronik, Naurtech, SAE Systems, Vangard Voice Systems, Voice-Insight, and Wavelink.

Unlike their predecessors, these new companies exploited the advancements made in commercial speech engines designed to be embedded in mobile devices. In parallel, they also took advantage of the COTS mobility devices: ruggedized small form-factor computers from companies such as Intermec, LXE, and Motorola (Symbol). While standardized mobile devices were still constrained in processing, memory and power, they were still highly capable of running more modern speech recognition algorithms.

Daniel Hong, lead analyst for Customer Interaction Technologies at Datamonitor, said, "These new commercial speech recognition offers bring a consistent attribute to the voice picking market: speaker-independent (SI) recognition technology, where the product works out-of-the-box without users having to train it to their specific voice. This eliminates many hours of lost staff time and speeds the time to ROI." All voice picking market entrants since 2002 have chosen the SI approach. In fact, no new SD vendor has entered the voice picking market with SD in the past eight years.

Speaker-independent (SI) technology is based on a broad set of acoustic models, allowing it to be highly insensitive to the manner in which a person speaks. Yet, like all speech recognition technologies, it may not

250

work for all speakers. As a result it is common to find adaptive speech approaches in SI technologies. At times the adaptation is manual in embedded technologies, where acoustic model sets can be appended from actual users.

Some of the pioneering vendors reacted to these new market dynamics. SyVox (acquired by Genesta in 2003) abandoned proprietary speech recognition and moved to commercial SI engines. Topsystem added an SI engine to its product line. Voxware moved away from being a hardware manufacturer to a software business model.

SOA and the Emergence of Thin Clients – 3rd Generation (2007+)

In the past decade and a half, many companies have embraced speech as a powerful Customer Service technology for use in their contact centers (as readers of this newsletter know). Unlike the fat-client recognition technologies discussed above in the first two generations of voice picking, contact center solutions have always deployed speech as a shared network resource. This topology only required users to place a phone call to access the speech recognition technologies. With the advent of Voice over IP support via WiFi networks (sometimes known as "VoFi"), a network-centric approach has been enabled for voice picking and other "inside the four walls" enterprise applications.

The new thin-client approach moves the speech recognition effort from each worker's mobile device to a centralized resource (server). This immediately reduces the processing and memory requirements on the mobile device; thus opening the door to support inexpensive devices, such as wireless IP telephones and smartphones. Handheld computers and PDAs can also be supported for multimodal transactions, as long as they support softphone capabilities (as many do).

Server-based speech recognition allows companies to tap into the most robust, open, and mature speech recognition technologies in today's markets – those that they have already been leveraging in their Customer Service contact centers. These speech recognition technologies are proven in today's market, handling billions of calls per day. They also excel at handling speech in noisy environments due to advancements made in the past 10 years to accommodate calls made over wireless networks and from noisy mobile environments.

Network speech solutions are speaker-independent (SI), bringing the "works right out of the box" and speedy deployment benefits discussed above. It is also common for network SI technology to be automatically adaptive, accepting new acoustic information into its database for improved performance based upon actual field use. Network speech recognition also supports the largest vocabularies (grammars) that are capable of understanding and acting on thousands of spoken words. This enables companies to use speech recognition resources beyond the warehouse.

W3C standards are also well-evolved for network-based SI technology, including a standard specification (MRCP) providing for the "plug-and-play" of speech recognition technologies from different vendors. This simplifies customer purchasing, as this new flexibility provides future-proofing, thereby eliminating being locked into a single speech recognition supplier. MRCP-compliant speech engines are available from companies making deep investments in speech technology, including Nuance, IBM, Microsoft, AT&T, and others.

"During any economic conditions, and particularly with the current global recession and credit crunch, Manufacturing and Supply Chain customers are intent to get more productivity and lifecycle services from their existing assets," says Chet Namboodri, Global Director for Manufacturing Industry Solutions at Cisco. "Thin-client solutions like warehouse voice picking that utilize existing wireless and converged IP infrastructure fall into this category. With incremental ROI paybacks at less than 12 months, customers can leverage their enterprise investment throughout the business with voice automation for any manual work flow, instead of proprietary point solutions."

The only vendor currently offering thin-client voice picking speech solutions is Datria, which brought it to market in 2007 in one of the industries largest voice picking deployments (2,325 concurrent users). That deployment enabled a large beverage company to use its Cisco wireless phones and speech recognition on Cisco routers to deploy 100 sites in a little over a year.

Technology Enablers

It would be disingenuous to attribute all technological advancements to the speech recognition software. That is far from the truth. All three speech recognition generations have benefited from dramatic improvements in noise-cancelling circuitry and directional microphones in commercially available headsets and mobile devices. Many handhelds are optimized for high quality audio transmission as well.

WiFi networks (802.11 a/b/g/n) have also evolved tremendously, providing stable and uninterrupted coverage inside warehouses and standardized levels of quality of service (QoS). VoIP support is now common on WiFi, easily engineered for roaming, shift-long connection times and 10+ hours on standard batteries.

Affordability has also been enhanced, as customers can now assemble voice picking solutions using commercial off-the-shelf mobile devices, headset, WiFi, and speech technologies.

Languages

At times, vendor marketing will focus on languages and dialects as another differentiator of SD and SI speech recognition technologies. This is a discussion that can mislead buyers. Multilingual capabilities imply two-way communications, both the ability to speak a language as well as hear/understand it. In most voice picking solutions – where text-to-speech synthesis can be a requirement to give voice to large catalogs of SKUs – it is the speaking to the worker that is the limiting factor. Most vendors use commercial text-to-speech (TTS) engines to synthesize speech in different languages, and those engines typically offer 10-25 dialect choices for embedded solutions and up to 35 for network-based solutions. This is far less than SI recognition (up to 58 dialects) or SD recognition (unlimited dialects). Notably, the number of language or dialect packs available tends to exceed the number of languages that supervisors and pickers actually speak in warehouses.

Language skills are exasperated in multimodal situations, where the ability to read a language (e.g., what is on a display) further complicates multilingual employee management.

One final anecdote about languages. Coca-Cola Enterprises deployed Spanish language support at some US locations due to the mix of employees speaking Spanish as their primary language. Unexpectedly, these employees chose to work in English, to build their second-language skills. To them, the voice picking solution acted as a language tutorial and provided a path to broadening their skill value to their employer.

Misrecognition

While this chapter has spent a considerable effort distinguishing pros and cons to varying speech-enabled warehousing technologies, it would be naïve to ignore that most speech recognition errors occur due to behavioral issues. Recognition issues commonly arise from users neglecting to:

Wear the headset microphone properly;

Speak in the expected manner (e.g.., yelling to overcome background noise instead of speaking in a normal voice;

Speak the expected words (being at a different prompt where a different set of words is expected); or

Use the right equipment (e.g., a headset with an omni-directional microphone).

It is incumbent on the voice picking vendor and its customer to jointly provide the necessary end-user training to avoid behaviors that result in misrecognition errors.

Summary – Standards transform markets

The voice picking market is following a natural market maturation curve familiar to many in IT. Early pioneers delivered vertically integrated solutions to establish that the concept worked and the business case was valid. Second-generation solutions moved towards openness, increasing hardware choices while reducing capital expenses and Total Cost of Ownership. The emerging third-generation of voice picking solutions completes the migration to 100% commercial-off-the-shelf components, W3C standards compliance, and unsurpassed choice, flexibility, and affordability. In addition, corporations can now choose voice picking solutions that create an enterprise-wide speech automation resource within their SOA architecture.

Today's diversity of speech recognition approaches is a boon to companies considering speech-enabled solutions for their supply chain.

Continuous Automated Speech Tuning and the Return of Statistical Grammars

Roberto Pieraccini, SpeechCycle

*In this chapter, Roberto Pieraccini, Chief Technology Officer, **SpeechCycle** suggests that, even in the most directed dialog interactions, a well-trained Statistical Language Model will easily outperform a rule-based grammar by allowing more varied responses, and discusses issues and solutions in using SLMs more widely. Roberto has been involved in spoken dialog technology for more than 25 years, both in research as well as in the development of commercial applications. Prior to joining SpeechCycle, Roberto was the manager of the Conversational Interaction Technology department at the **IBM** Thomas J. Watson Research Center. Before that, he led the Natural Dialog R&D group at **SpeechWorks International** (now Nuance). Earlier, he joined the Speech Research Group at **AT&T** Bell Labs and later AT&T Shannon Labs. Roberto began his career as a speech scientist with **CSELT**, the research center of the then Italian operating telephone company, after completing his doctorate in engineering from the **Universita degli Studi di Pisa**, Italy.*

Statistical grammars, commonly dubbed "SLMs" (for Statistical Language Models) by IVR practitioners, have been known to the speech research world for almost 30 years. However SLMs started to make their first steps into the IVR world only relatively recently. On the other hand, rule-based grammars, often written in an XML dialect known as SRGS (Speech Recognition Grammar Specification) have been used since the first attempts to deploy speech recognition in the early 1990s. Rule-based grammars are king for directed dialog, and only for situations where an open prompt[10] needs to be played, are SLMs brought into the equation with a lot of effort and mystique. But what the IVR world often ignores is that, even in the most directed dialog interactions, say plain YES/NO questions, a well-trained SLM will easily outperform a rule-based grammar, all other things being equal.

So, why are we using handcrafted, rule-based, XML grammars at all if we know that SLMs would work better? There are several reasons for that. First, building an SLM is not as easy as writing rules in XML. You need data, and it is the kind of data that you don't have when you first build a new spoken dialog system. And even if you had data—I mean recorded utterances as responses to each prompt—you would need it transcribed. On top of that you would need to provide for each

---

[10] AT&T's "How may I help you?" is the classic example of such an open prompt.

transcribed training utterance—in a process called *annotation*—a semantic tag: its meaning. So while instructing transcribers to do the right thing can be straightforward, teaching annotators the correct utterance-tag mappings may be very challenging. And assuming you have your transcribed and annotated utterances, and you have cleaned all of the transcriptions and annotations to transform them into a consistent set of training data—and you need lots of them—now you have to build an SLM. How do you build an SLM?

Building an SLM requires you to have a special set of programs that do two things. The first is creating a properly called[11] *statistical language model,* in other words a way to tell the recognizer which sequences of words are legal. While rule-based grammars explicitly list all the legal sequences of words, a statistical language model does it in a …well…statistical sense [1]. This is accomplished by computing what people of the trade call *n-grams,* which are the probabilities for any possible word—at least for all the words that appear in the training utterances—to be preceded by any possible *n-1* long sequence of words. So, if *n=3,* as it typically is in practice, the statistical language model computes, for each word, the probability of being preceded by all the possible sequences of 2 words. Thus, if you had 1,000 words in your vocabulary—and typically you need more than that for an open prompt, less than that for a directed dialog prompt—the program that computes the statistical language model has to compute 1000 (all the possible words) times 1000x1000 (all the possible pairs of words) probabilities, in other words 1 billion probabilities! Don't worry … you may not be able to find examples of all the possible 1 billion triplets of words of your measly 1,000-word vocabulary even if you search the whole Web. But the statistical language model training program has to provide a number, the estimate of a probability, even for the most unlikely triples, like *yes maybe computer*[12]. But again, don't worry. There are programs that do that for you, and you can buy those programs; you can get them for free from some open-source packages published on the Web, or if you are versatile

---

[11] Although the *statistical language model* is only a part of an SLM, the industry term SLM (Statistical Language Model) indicates the full ability to decode the meaning out of free-form utterances, or *natural language* utterance, typically the responses to an open prompt.

[12] In fact, when an uncommon triplet of words is not found in the training set, statistical language models approximate that probability using some heuristic considerations. There should not be any triplet of words with a zero probability, since that would a-priori exclude that triplet, however uncommon, for being ever recognized.

enough and not afraid of a little math and some algorithms, you can build them yourself. But of course, even if you had the best of programs for building the best statistical language model, you have to fiddle with a number of parameters in order to get the best out of it. And that may not be easy. But that's not all.

Building an SLM is not just about constraining the recognizer on all the possible sequences of words. In a spoken dialog system you don't need just the words that were spoken by the caller, but a semantic tag, a symbolic output from a set of *slots* that has a meaning for the call-flow at the particular prompt. For instance, if the prompt is asking a simple YES/NO question—for example "Have you paid your most recent bill?"—you want the recognizer to return either a YES or a NO. If the caller says *yes*, you want the recognizer to return YES; if the caller says "you bet" you want the recognizer to return YES; if the caller says "no way" you want the recognizer to return NO, and so on. If your training utterances are semantically annotated, on top of being transcribed, SLM builders create what is called a *semantic classifier* [2], in other words a program that takes as input the string of words recognized by the speech recognizer, and returns one out of a number of slot identifiers. Semantic classifiers are built from large samples of transcribed and annotated utterances. Again, you can buy one of these programs, or you can get it from some open-source projects, or if you feel adventurous in some non-trivial math and some non-trivial algorithms then you can build it yourself. And even if you buy it, you still have to adjust parameters and do some non-trivial tuning if you want to get the best performance.

In short, this lack of data and expertise has made using of SLMs unpopular especially for directed dialog solutions since the early days of speech IVR technology. Think about how much easier building an XML rule-based grammar is in comparison.

But there is another reason for SLM's lack of popularity. If you create a rule-based grammar, you can see what you did. You can immediately understand why the recognizer did not recognize the phrase "maybe yes" spoken by an undecided caller—perhaps because that phrase was not in the grammar—and you can promptly make the necessary modifications in a matter of minutes. You cannot easily do that with statistical grammars. N-grams, probabilities, and statistical classifiers are inscrutable at first sight. If something goes wrong, you need some understanding of the statistical machine learning theory behind the SLM in order to fix it. Also, the common notion—common and widely accepted in research as a logical and experimental fact—that SLMs always outperform rule-based grammars if trained on the right data, is not wholeheartedly embraced by IVR practitioners. There is confusion between the performance of an SLM in an open prompt situation, and the performance of a rule-based grammar in directed dialog. Of course the latter works better than the former, but only assuming callers always say what is in grammar, which is not always true. But this notion muddles the notion that a properly trained SLM will outperform a rule-based grammar in a directed dialog situation, especially when callers say things that are out of grammar. Yes, a well-trained SLM will have at least the same performance, and most likely outperform a corresponding rule-based grammar. So, why are we using rule-based grammars at all? As the previous paragraphs elucidate, it has nothing to do with grammar performance and everything to do with the high price of admission that SLMs entail.

And here comes the idea. What if we could provide a way to create and tune statistical grammars *automatically*, and use them for every context in a dialog, either open-prompt or directed, in place of traditionally handcrafted rule-based grammars, and do that continuously, while the application is deployed? After all, except for transcription and annotation, there is nothing that strictly requires the continuous labor of machine learning and speech scientists to build SLMs, while that is not always true for rule-based grammars. Yes, speech scientists run experiments to determine the best set of parameters, but they can create programs that run the experiments for them and decide which best selection of parameters to pick. Yes, they condition and clean the data, making sure that there are no inconsistencies in the transcriptions and annotations. But they, the machine learning and statistic speech experts, can create programs that do that for them. They can also create programs that help reduce the cost of human transcription and annotation by automating it when possible. Experienced speech scientists who are also computer scientists and machine learning experts can work on programs that tune speech grammars, rather than working on speech grammars themselves. And programs, unlike humans, can handle an abundance of data, and the speech grammars can get better and better.

What I described is the concept of automated tuning [3], or *grammar factory*, which is a service that can take a constant flow of log data from your IVR and give you, continuously, better and better grammars. How much better? Well, that depends on other factors, such as the design of the call flow, the prompts, the task, and other things that can give rise to poor grammar performance if not properly done. But while the grammar factory is looking at improving grammars, it can also flag behavior that would require the inspection of expert VUI designers and speech scientists. Will the grammars keep improving indefinitely? Certainly not. After a while, after enough data has been processed, their performance will settle on the maximum possible performance, which may not be 100% accuracy because of all of the other factors that can influence speech performance. But one thing is sure. It would be very hard, if not impossible, to reach that theoretical maximum performance using old-fashioned, handcrafted grammars tuned by hand. And if any change occurs in the application—a new prompt, a new strategy, new products, a new language—the grammar factory will, automatically and relentlessly, adjust for that and guarantee, in a short time, the attainment of the best performance for the speech recognizer. This is a new step ahead in the direction of machines that truly understand speech and continuously learn from what they hear.

References

[1] Young, S., "Talking to Machines (Statistically Speaking)," Proceedings of the 7th International Conference on Spoken Language Processing (ICSLP 2002), September 16-20, 2002, Denver, Colorado.

[2] Evanini, K., Suendermann, D., Pieraccini, R., "Call Classification for Automated Troubleshooting on Large Corpora," Proceedings of the 2007 IEEE ASRU Workshop, Kyoto, Japan, December 9-13, 2007

[3] Suendermann, D., Evanini, K., Liscombe, J., Hunter, P, Dayanidhi, K., Pieraccini, R., "From Rule-Based to Statistical Grammars: Continuous Improvement of Large-Scale Spoken Dialog Systems," Proceedings of the 2009 IEEE Conference on Acoustics, Speech and Signal Processing (ICASSP 2009), Taipei, Taiwan, April 19-24, 2009

In closing
William Meisel

These chapters reflect the expertise of many talented individuals in the use of speech technology in the user interface. The desire to find what works for users is clear in every chapter.

Earlier speech application pioneers had to search for ways to overcome the limits of the technology. Some applications even cautioned users that they should talk to speech recognition systems as if they were talking to their dog. Developers are far from that world today.

Speech technology will continue to improve in its capabilities, if for no other reason than the computing power and speech data collection required to refine the technology and deliver it will become increasingly cheaper. And we can hope for core technology improvements as well, perhaps aided by advances in natural language processing and interpretation. We can increasingly concentrate on what helps users, rather than technology limitations. This book has been about that goal.